SCIENTIFIC
FEUDS

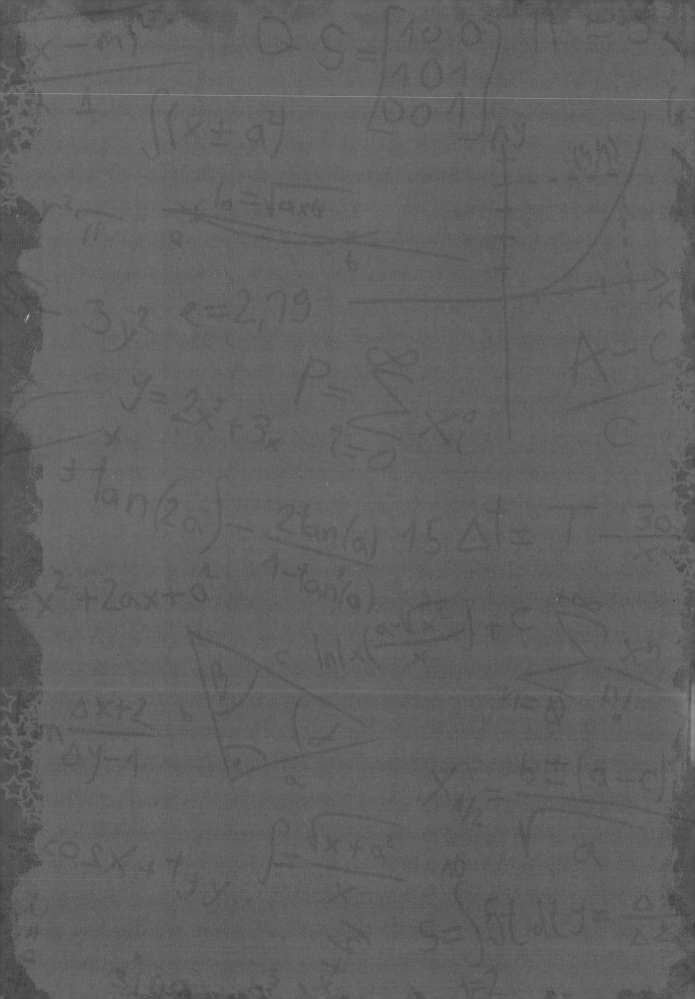

SCIENTIFIC FEUDS

JOEL LEVY

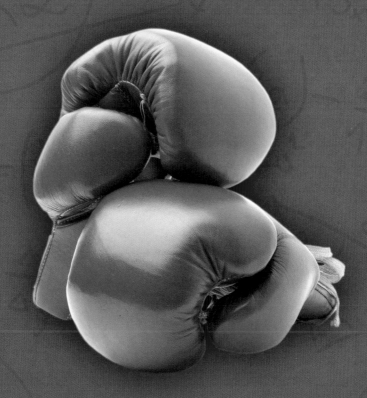

NEW
HOLLAND

For Matt, Aisling and Sam

Published in 2010 by New Holland Publishers (UK) Ltd
London • Cape Town • Sydney • Auckland
www.newhollandpublishers.com

Garfield House, 86–88 Edgware Road, London W2 2EA, United Kingdom
80 McKenzie Street, Cape Town 8001, South Africa
Unit 1, 66 Gibbes Street, Chatswood, NSW 2067, Australia
218 Lake Road, Northcote, Auckland, New Zealand

10 9 8 7 6 5 4 3 2 1

A catalogue record for this book is available from the British Library

Hardback ISBN 978 1 84773 514 0
Paperback ISBN 978 1 84773 717 5

Senior Editor: Kate Parker
Editorial Direction: Rosemary Wilkinson
Publisher: Aruna Vasudevan
Design and cover design: David Etherington
Illustrator: Alan Marshall
Production: Melanie Dowland

Reproduction by Pica Digital PTE Ltd, Singapore
Printed and bound by Tien Wah Press, Singapore

▶ **Fight the power.** *Italian fresco of the 13th century showing Galen and Hippocrates, the fathers of medicine. For over a millennium their authority on matters medical was unchallenged, but in the 16th and 17th centuries a few brave individuals triggered controversy – and bitter feuding – by daring to point out their mistakes.*

CONTENTS

FEATURES

INTRODUCTION

T he history of science is boring; the traditional version, that is, with its stately progression of breakthroughs and discoveries, inspirational geniuses and the long march out of the darkness of ignorance into the light of knowledge. This is the story as it is often presented in museums, textbooks and classrooms; but it is an invention, a typically Victorian piece of bowdlerized mythmaking. The real history of science is far messier, more nuanced and complex, and much, much dirtier. This was true of the earliest scientists, and it is still true today. According to Nobel prize-winning physicist Leon Lederman, speaking in 1999: 'You'd think that scientists would have a degree of saintliness that would be almost unbearable. It doesn't work that way. The competition goes on at all levels – the international, the national, the institutional, and finally the guy across the hall.'

The ancients, too, knew that conflict was inevitable in science. The father of medicine, Hippocrates, in his 4th-century BCE work, *The Law*, wrote that, 'There are in fact two things, science and opinion; the former begets knowledge, the latter ignorance', while according to Roman naturalist Pliny the Elder, writing in the first century CE, 'This only is certain, that there is nothing certain; and nothing more miserable and yet more arrogant than man.'

This book surveys more than 25 feuds from the history of science and technology, from the very beginnings of science in the Early Modern Period (roughly 1500–1750) to recent feuds in areas such as genomics and human evolution. Some feuds were little more than cordial disagreements over points of theory but many were vicious and prolonged, and some, all-consuming. The stakes were often great – eternal glory, Nobel prizes, untold wealth, personal ruin, even life itself – and the details often unedifying. Collateral damage ranged from blighted careers to electrocuted elephants. The range and variety of the feuds makes it hard to generalize, but each story is revealing in its own way: about key scientific debates, but also about how science works.

▲ **Face off.** An engraving after Joseph Nicolas Robert-Fleury's 1847 painting, Galileo Before the Inquisition. Galileo's scientific ideas were attacked by many, but it was only when he inadvertently triggered a feud with his former patron Maffeo Barberini, aka Pope Urban VIII, that he found himself on trial for his life.

The feudal system

Science was characterized by feuding from the very beginning. Its roots were in alchemy and magic, and while some alchemists and magi collaborated, in the main they toiled in isolation, guarding their secrets, denigrating the efforts of others and hoping that they alone would be the first to achieve the ultimate prize – the Philosopher's Stone, transmutation of lead into gold, the Elixir of Life, the restitution of ancient wisdom. As natural philosophy took its first steps towards the scientific world view with the astronomical discoveries of Copernicus, Kepler and Galileo, relations between the great men of the day were as often defined by fear and loathing as respect and admiration. Johannes Kepler served as an assistant to the great Danish astronomer Tycho Brahe, but the two fell out at least once in the short time they knew each other, while Tycho also carried on a bitter feud with rival astronomer Ursus (*see* pages 164–165) and was cordially despised by Galileo.

Matters did not improve when the baton of natural philosophy passed to England in the late 17th century, where Sir Isaac Newton, perhaps the greatest scientist of all time, was also perhaps the most argumentative. Even his famous phrase, 'If I have seen further it is by standing on the shoulders of giants', which appears to encapsulate everything that is humble, noble and gracious about science, may have been little more than a dig at a short-statured enemy (*see* pages 180–181). Why is science such a dirty business, and what does this long history of dispute say about the nature of science?

Foibles and mistakes

Scientists are people too, although this is easily forgotten thanks to the way science is typically portrayed in the media (*see* pages 136–139). The work they do is carried out in a social context like any other human endeavour. Scientific ideas themselves emerge from and are often representative of this social context. While the Victorian historiography of science tended to obscure these simple facts, more recent historians of science appreciate that it can only be understood in the light of them. 'We have come a long way from acceptance of the conventional Victorian belief in the disinterested scientist engaged in the objective pursuit of Truth,' comments Tony Hallam, a professor of geology and palaeontology (historically two of the more contentious sciences), 'to a less lofty but more realistic one which takes account of the existence of a whole range of social interactions within the scientific community as determinants of scientific theory.'

Naturally, then, science is prey to the same foibles, insecurities and mistakes as any other field. Just as in politics or sport, there will be personality clashes and power grabs, misunderstandings and betrayals; and, just as in these fields, the people who rise to the top, who achieve the most, are likely to be driven and bloody-minded. As astrophysicist Virginia Trimble points out: 'nobody who does something earthshaking is likely to be easy to get along with. You only achieve things like that by being more single-minded than your friends and relations are likely to regard as totally reasonable.'

Nasty, brutish and smart

Above and beyond this, however, the nature of science means that conflict is built into its DNA. Science in its purest form is a process of trial and error: hypotheses are formed through observation and experiment, and then these hypotheses are tested with further observation and experiment. If they are supported, they become theories – 'true' models of how the world works, perhaps even laws of nature – but even the most solid theory can be revised or overturned if new evidence comes to light (*see* pages 98–101). This ideal of the scientific method has led some theorists of science to apply Darwinian ideas of natural selection to science itself: ideas are engaged in a constant battle for survival, in which only the fittest will prosper. If science really is so combative by its very nature, it is only to be expected that conflict will result. When the natural proclivities of driven, single-minded individuals are added to this, a combustible mixture results.

Scientists are defined by their ideas and entire careers can hang on a theory, model or interpretation; inevitably they will fight their corners and oppose those who hold competing ideas. Modern science introduces many exacerbating factors – the scramble for funding, the imperative to publish, the politics of academia. Perhaps feuding is the default state for science, and instances of collaboration and concord the real curios.

▶ **British bulldog.** *A caricature of T.H. Huxley from the January 1871 issue of* Vanity Fair. *Huxley was one of the most pugnacious scientists of his era, and delighted in fighting Darwin's battles for him, earning Huxley the nickname 'Darwin's bulldog'.*

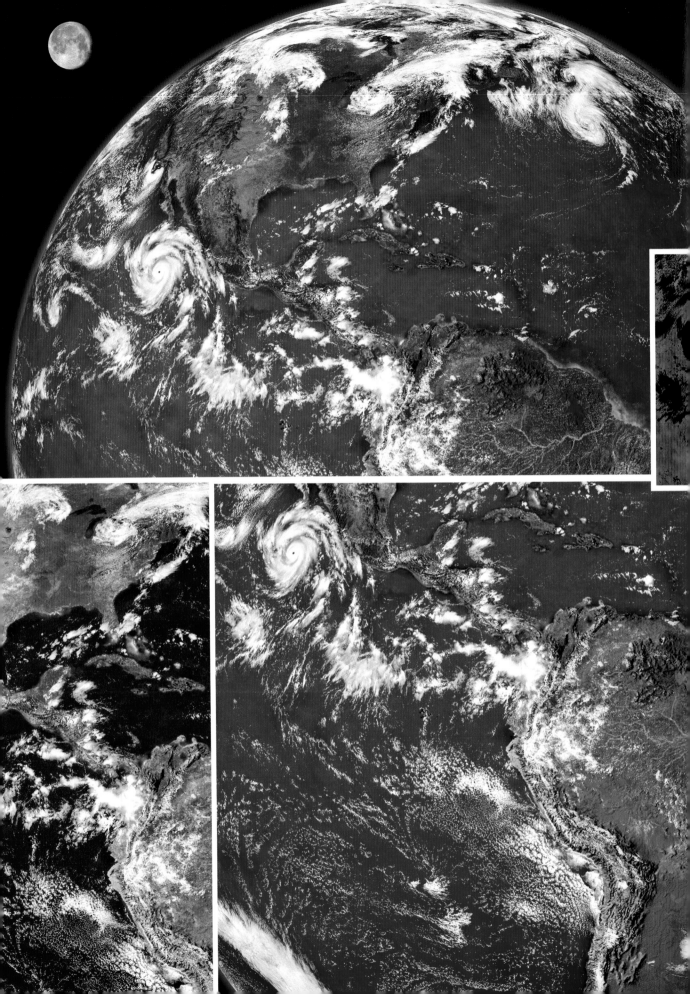

PART ONE
EARTH SCIENCES

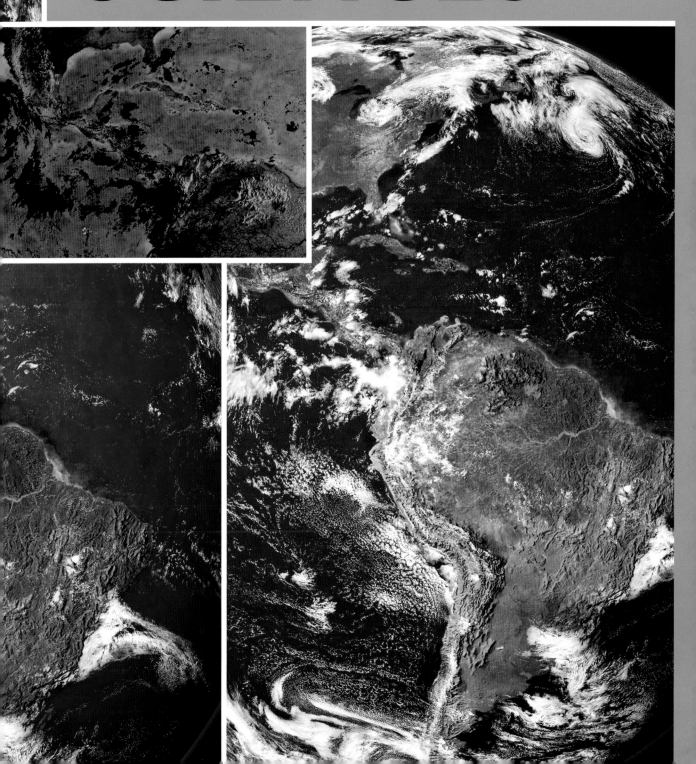

KELVIN
vs
LYELL, DARWIN, HUXLEY, et al.

FEUDING PARTIES
William Thomson, Lord Kelvin
(1824–1907) – physicist, grand
old man of British science
vs
Sir Charles Lyell (1797–1875)
– geologist;
Charles Darwin (1809–82)
– naturalist;
T.H. Huxley (1825–95)
– biologist;
and many others

DATE
1861–1904

CAUSE OF FEUD
Debate over the age of the Earth

Early scientists, including Newton, generally believed that the biblical account of the Creation was literally true, and therefore that the internal chronology of the Bible could be used to calculate the age of the Earth. Newton himself arrived at a figure of around 6,000 years but it was the Anglo–Irish Archbishop of Armagh, James Ussher, who, in a feat of formidable scholarship, determined that Creation began in the early hours of Sunday, 23 October 4004 BCE.

'Incomprehensibly vast'

Ussher's 1654 calculation remained the mainstream view until the 18th century saw the birth of a new science, geology – the study of how the earth was shaped. It became obvious to the practitioners of this nascent science that the processes and phenomena they observed must be acting on timescales much larger than the few millennia allowed by biblical literalism. The deposition of rocks, the uplift and folding of strata and mountains, the erosion of valleys and cliffs; all these spoke of slow processes operating over long periods. The emerging evidence of fossils, with their record of strange forms now vanished from the Earth, also suggested a long passage of time. Indeed, to geologists such as Charles Lyell, author of the seminal *Principles of Geology*, it seemed likely that natural processes of rock formation and erosion had been occurring for an effectively incalculable length of time; if not for a limitless period, certainly of the order of billions of years.

Meanwhile, another group of scientists was approaching the problem of the age of the Earth from a different angle. Naturalists were becoming increasingly convinced that species of plants and animals had changed over time through some form of evolution, but that this transformative process operated extremely slowly, and therefore constituted its own brand of evidence for the great age of the planet. The expanses of geological time opened up by Lyell were a key plank of Darwin's argument in his 1859 publication *On the Origin of Species*, in which he cautioned: 'He who can read Sir Charles Lyell's grand work on the *Principles of Geology* and yet does not admit how incomprehensibly vast have been the past periods of time, may at once close this volume.' To illustrate just how vast these periods had been, Darwin included a rough estimate he had made of the length of time it must have taken for the ocean to erode the Weald (a geological feature in the south-east of England), putting it at around 300 million years.

Lord Kelvin objects

To many at the time, such immense numbers seemed equivalent to eternity, and Lyell and, by extension, Darwin were seen as the standard-bearers of a school of geological thought called uniformitarianism. In its most extreme form, the uniformitarian view was that the Earth had effectively existed forever, and might well continue to do so, its geological processes endlessly cycling through the creation and destruction of landscape features. Lyell and Darwin did not subscribe to this extreme view, but they nonetheless became targets of the ire of a man who had proved that this theory of a never-ending cycle was impossible.

William Thomson, elevated to the peerage as Baron Kelvin of Largs in 1892 (the first scientist to be so honoured) and hence conventionally referred to as 'Lord Kelvin' or 'Kelvin', had elucidated, among other achievements, the laws of thermodynamics. Briefly stated, these meant that new energy could not be created out of nothing, and that the energy of any system would tend to dissipate. The laws meant that a perpetual-motion machine was impossible, and the endlessly recycling and eternal Earth of the extreme uniformitarians was effectively just that. Kelvin was having none of it.

CHRONOLOGY.

'Bus-Driver. "They tell me there've been some coins found in these 'ere 'exkyvations' that 'a been buried there a matter o' four or five 'undred year!!"

Passenger Friend. "Oh, that's nothin'! Why, there's some in the Bri'sh Museum—ah—more than Two Thousand Year Old!!"

'Bus-Driver (after a pause). "Come, George, that won't do, yer know! 'Cause we're only in Eight'n 'Undred an' Sixty-Nine now!!!"

◀ *Chronological confusions.* A cartoon from the satirical magazine *Punch*, from 1869, lampooning the use of the Bible as a basis for determining the age of the Earth.

In March 1862, Kelvin published a paper, 'On the age of the sun's heat', in which he calculated that the Sun had been burning for less than a million years. 'What then,' he asked, 'are we to think of such geological estimates as 300,000,000 years for the "denudation of the Weald"?' Given that his estimate of the age of the Sun, though imprecise, was based on 'known physical laws' and was orders of magnitude less than the figure arrived at by Darwin, he suggested that it was probable that the naturalist had underestimated the speed of erosion that could be caused by 'a stormy sea, with possibly channel tides of extreme violence'.

Kelvin was one of the world's great authorities on the dynamics of heat. He started with several assumptions: that the Earth had begun as a ball of molten rock; that, in accordance with his laws of thermodynamics, no heat could have been added to the system since this formation; and that convection currents had allowed uniform cooling of this molten globe to a solid sphere of uniform temperature, which then radiated the rest of its heat out into space from its surface. It was widely known from mining that the ground got hotter as you went down, with a thermal gradient of about 1°F per 50 feet (0.5°C per 15 metres). Kelvin did his own experiments to determine the thermal conductivity of rocks, and employed his mastery of Fourier mathematics to work out how long it must have taken for the planet to cool to its current temperature. He arrived at an estimate of 98 million years, with a range of 20–400 million years for the highest and lowest possible figures.

Burned fingers

Kelvin's calculation carried enormous authority, thanks both to his eminence and to the manner in which he seemed to have applied 'hard' science and 'pure' mathematics to a field that had previously been the victim of woolly thinking. The mature and respectable science of physics had set straight the immature new discipline of geology. Darwin was chastened; he considered the age limit that Kelvin had placed on the Earth to be the most serious and credible argument against his carefully worked out theory, which demanded immense epochs of time. The scientist Fleeming Jenkin summarized the argument: 'The estimates of geologists must yield before the more accurate methods of computation, and these show that our world cannot have been habitable for more than an infinitely insufficient period for the execution of the Darwinian transmutation.' Darwin himself referred to Kelvin as his 'sorest trouble' and an 'odious spectre'.

TIMELINE

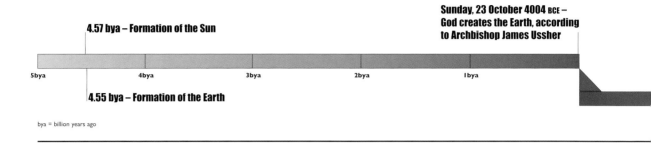

4.57 bya – Formation of the Sun

Sunday, 23 October 4004 BCE –
God creates the Earth, according
to Archbishop James Ussher

5bya   4bya   3bya   2bya   1bya

4.55 bya – Formation of the Earth

bya = billion years ago

Darwin was so troubled that in the third edition of *On The Origin of Species* he removed his Weald calculation altogether, but even this did not quiet the criticism from Kelvin and his allies. In April 1869, Darwin was moved to warn Lyell theatrically: 'Having burned my fingers so consumedly with the Wealden, I am fearful for you … for heaven's sake take care of your fingers: to burn them severely, as I have done, is very unpleasant.'

'The grandest mill'

Darwin may have been running scared, but his self-appointed bulldog, T.H. (Thomas Henry) Huxley (*see* pages 38–47), was ever ready to take up the gauntlet on his behalf. In his 1869 presidential address to the Geological Society of London, Huxley defended the views of the 'old Earthers' and pointed out the basic flaw in Kelvin's approach: 'Mathematics may be compared to a mill of exquisite workmanship, which grinds your stuff to any degree of fineness; but, nevertheless, what you get out depends on what you put in; and as the grandest mill in the world will not extract wheat flour from peas cods, so pages of formulae will not get a definite result out of loose data.' In other words, Kelvin's calculations might be unimpeachable, but if he had got his starting assumptions wrong then his conclusions would also be wrong.

The scientific world now began to gather behind the respective banners of Kelvin and Huxley. And although the physicist P.G. Tait used a new method to calculate that the Sun was around 20 million years old and the Earth only 10 million, in the tenth edition of his *Principles of Geology*, Lyell accepted that the age of the Earth was finite but dated the Cambrian era to around 240 million years ago. Although many scientists sought some middle ground, attempting to prove that evolutionary and geological processes might act relatively quickly, operating within Kelvin's timescale, it was becoming increasingly obvious that both sides could not be correct; someone must be making basic errors.

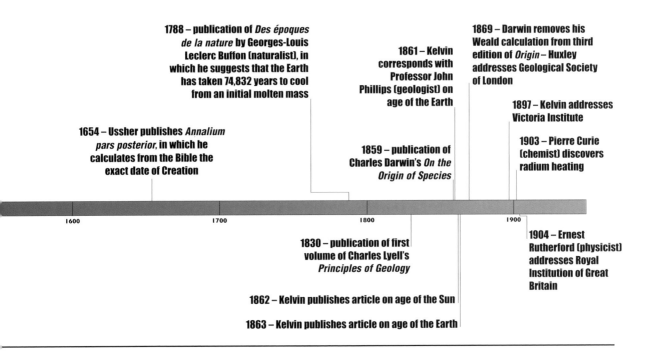

1654 – Ussher publishes *Annalium pars posterior*, in which he calculates from the Bible the exact date of Creation

1788 – publication of *Des époques de la nature* by Georges-Louis Leclerc Buffon (naturalist), in which he suggests that the Earth has taken 74,832 years to cool from an initial molten mass

1861 – Kelvin corresponds with Professor John Phillips (geologist) on age of the Earth

1859 – publication of Charles Darwin's *On the Origin of Species*

1869 – Darwin removes his Weald calculation from third edition of *Origin* – Huxley addresses Geological Society of London

1897 – Kelvin addresses Victoria Institute

1903 – Pierre Curie (chemist) discovers radium heating

1830 – publication of first volume of Charles Lyell's *Principles of Geology*

1904 – Ernest Rutherford (physicist) addresses Royal Institution of Great Britain

1862 – Kelvin publishes article on age of the Sun

1863 – Kelvin publishes article on age of the Earth

The geophysicist Osmond Fisher suggested that the error was Kelvin's, proposing a new (and prescient) model of the structure of the Earth – a thin crust over a plastic substratum – that would destroy the basic assumptions upon which Kelvin had based his calculations. Fisher further pointed out that it was a form of scientific arrogance to disregard the clear evidence of geology and biology: 'I think we cannot but lament, that mathematical physicists seem to ignore the phenomena upon which our science founds its conclusions, and, instead of seeking for admissible hypotheses the outcome of which, when submitted to calculation, might agree with the facts of geology, they assume one which is suited to the exigencies of some powerful methods of analysis, and having obtained their result, on the strength of it bid bewildered geologists to disbelieve the evidence of their senses.'

In seeking to approach the question of the age of the Earth from the point of view of 'proper scientists', the mathematical physicists were actually betraying one of the cardinal rules of science: if the facts do not fit the theory, the theory must be modified or discarded, not the other way round.

The fire within

With opposition to his views mounting, Kelvin was stung to respond. In 1897, he addressed the Victoria Institute with a talk entitled 'The age of the Earth as an abode fitted for life'. Many assumed he would modify his views or relent. In fact, he was more

◄ The Badlands of Dakota. This was once the bed of an inland sea. Layers of rock have been revealed by aeons of gradual erosion – or perhaps by a Biblical deluge?

dogmatic and intransigent than ever, revising his uppermost estimate of the age of the Earth to 24 million years, and talking grandly of 'certain truths'.

Unfortunately for Kelvin, new discoveries were at hand that would invalidate his fundamental assumption that the Earth had started with limited heat energy and lost heat ever since. Radioactivity had been discovered in 1896, and in 1903 the French chemist Pierre Curie realized the potential geological significance of heat generation by radioactive elements in rocks. The following year, the greatest of the new generation of physicists, Ernest Rutherford, lectured at the Royal Institution of Great Britain on the topic of radium and of radioactive elements as a source of heat energy. Noticing Lord Kelvin in the audience, he realized he was 'in for trouble at the last part of my speech dealing with the age of the Earth, where my views conflicted with his. To my relief, Kelvin fell fast asleep, but as I came to the important point, I saw the old bird sit up, open an eye and cock a baleful glance at me! Then a sudden inspiration came, and I said Lord Kelvin had limited the age of the Earth *provided no new source of heat was discovered.* That prophetic utterance refers to what we are now considering tonight, radium! Behold, the old boy beamed upon me.'

In practice, Kelvin continued to deny that radioactivity had rewritten the rules of the debate. However, by the time he died in 1907, radioisotope dating was already being used to make direct measurements of the age of rocks, with samples dated at up to 2.2 billion

> *'As Lord Kelvin is the highest authority in science now living, I think we must yield to him and accept his views.'*
>
> MARK TWAIN, LETTERS FROM THE EARTH, 1909

years old. By 1931, the geologist Arthur Holmes was able to assure a US National Research Council meeting that 'the age of the Earth exceeds 1,460 million years [and] is probably not less than 1,600 million years'. Modern dating techniques reliably prove that the Earth is around 4.55 billion years old (*see* box below).

EVIDENCE FOR THE AGE OF THE EARTH
How do we know how old the Earth is?

There are three primary methods, all of which measure the age of rock by comparing the ratios of isotopes they contain (radiometric dating). The oldest rocks so far discovered on Earth are around 3.9 billion years old, and some include minerals that are even older (around 4.2 billion years old). This puts a lower limit on the age of the Earth, but not an upper one, since none of the original surface of the Earth still exists, thanks to its molten nature and subsequent processes of erosion and crustal recycling.

More direct means of calculating the age of the Earth are based on the assumption that all the rocky material in the solar system formed at the same time, and from the same pool of material (as a giant disc of dust and gas coalesced into solid matter). Different isotopes of uranium decay into different isotopes of lead and, by measuring the ratios of these isotopes in Earth rock and meteorites, it is possible to plot a graph of the values and calculate from it the amount of time that has elapsed since the original pool of matter became separated into discrete objects. This method, known as lead isochron dating, gives a figure of around 4.55 billion years, as does the other direct method, which is radiometric dating of meteorites (asteroids that have fallen to Earth). Unlike the Earth, asteroids do not undergo geological processes and therefore may date back to the formation of the Solar System. Around 100 meteorites have been dated, and the ages obtained are almost all around 4.5 billion years.

WEGENER

vs

JEFFREYS, et al.

FEUDING PARTIES
Alfred Lothar Wegener
(1880–1930) – German
meteorologist and astronomer,
architect of continental drift
theory
vs
Sir Harold Jeffreys (1891–1989)
– geophysicist;
most of the rest of the
geological world

DATE
1915–60s

CAUSE OF FEUD
The theory of continental drift

The story of Alfred Wegener and the theory of continental drift is often cited as a prime example of how theories that are beyond the pale can rapidly become accepted scientific dogma and of the way in which the real narrative of science (messy, contradictory and contentious) is quite different from the 'classical' picture of a serene progress from ignorance to enlightenment. In truth, the Wegener case is not a brilliant illustration of either of these, but it remains a popular and fascinating story.

The shrinking Earth

The most obvious evidence for the notion that the continents were once joined is the apparent fit between the coastlines of eastern South America and western Africa, which became apparent almost as soon as the first maps of the New World were produced. As early as 1596, Dutch map-maker Abraham Ortelius suggested that the Americas had once been joined to Europe and Africa, until they had been 'torn away … by earthquakes and floods'.

In 1881, pioneer geophysicist Osmond Fisher proposed a model of the Earth in which a crust of hard rock sat atop a fluid mantle, even suggesting that the ocean floor expanded through volcanic production of new rock, and that contraction of the continents gave rise to mountain ranges. This was a remarkably prescient prototype of modern plate tectonics theory, which went unheralded. Fisher was going against the grain of an Anglo–American tradition that emphasized the relative permanence of the oceans and continents. While in the German-speaking world the notion that the crust of the planet was mobile and the interior fluid had some currency, the mainstream hypothesis was that the Earth was cooling from an initial molten state (*see* page 17), and in the process contracting, so that its skin rumpled and creased, creating mountain ranges and oceanic basins.

Scientific adventurer

Onto this scene burst Alfred Wegener, an intrepid meteorologist and astronomer. Struck, like many, by the jigsaw-like fit of the South American and African coastlines, Wegener was intrigued when, in 1911, he came across a report outlining palaeontological connections

TIMELINE

*c.*190 mya – Pangaea
breaks up into
Gondwana and other
land masses

16th century – first maps of
the New World

1500

*c.*250 mya – Pangaea
supercontinent

300mya   200mya   100mya

mya = million years ago

between Brazil and Africa (such as fossils of the same species on both continents). Many other such connections between far-flung regions were known, but they were generally presumed to indicate the former existence of land bridges, now sunk beneath the oceans.

Wegener claimed not to be aware of the 'continental drift' hypothesis of the American geographer F.B. Taylor, published in 1910, and in 1912 he came up with his own, similar theory. He fleshed out this theory in a 1915 book, *Die Entstehung der Kontinente und Ozeane* (*The Origin of the Continents and Oceans*), in which he laid out several strands of evidence for his theory of '*Die Verschiebung der Kontinente*', properly translated as 'continental displacement'.

◀ **Meteorological expedition in the Arctic.** Wegener's research background was in meteorology rather than geology. The vitriolic response to his geological theories was partly motivated by his outsider status.

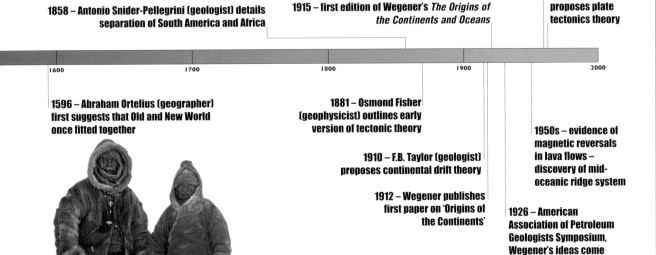

1960 – Harry Hess (geologist) proposes 'spreading sea-floor' hypothesis

1965 – Tuzo Wilson (geophysicist) proposes plate tectonics theory

1858 – Antonio Snider-Pellegrini (geologist) details separation of South America and Africa

1915 – first edition of Wegener's *The Origins of the Continents and Oceans*

| 1600 | 1700 | 1800 | 1900 | 2000 |

1596 – Abraham Ortelius (geographer) first suggests that Old and New World once fitted together

1881 – Osmond Fisher (geophysicist) outlines early version of tectonic theory

1950s – evidence of magnetic reversals in lava flows – discovery of mid-oceanic ridge system

1910 – F.B. Taylor (geologist) proposes continental drift theory

1912 – Wegener publishes first paper on 'Origins of the Continents'

1926 – American Association of Petroleum Geologists Symposium, Wegener's ideas come under attack

◀ **Wegener, left,** on one of his research trips in the Arctic.

Wegener's theory

Wegener began by discussing flaws in the current contraction model: for instance, if the globe was uniformly contracting, why were the mountain ranges and ocean basins so unevenly distributed? He pointed to clear evidence that there were two distinct types of crustal rock – continental and oceanic – and tried to show how the strata underlying continental crust could deform under great pressure over long periods of time until it acted almost like a fluid (much as ice will). He collated evidence of similarities in rock types and strata on either side of the Atlantic, which suggested former contiguity, arguing: 'It is just as if we were to refit the torn pieces of a newspaper by matching their edges and then check whether the lines of print run smoothly across.'

To this evidence he added the mounting evidence from the fossil record of species found on both sides of the Atlantic, such as Mesosaurus, a small reptile from the Permian era, and Glossopteris, a plant from the Permo-Carboniferous era. These distributions could not be explained by now-sunken land bridges, he pointed out, because such land bridges were an impossibility: continental granite was less dense than oceanic basalt, and therefore could not sink into the ocean floor. The prevailing belief in these land bridges was, he wrote, 'a perfectly preposterous attitude'.

Wegener was particularly impressed by the presence at high latitudes of rock types and coal deposits that must have formed in the tropics. All this evidence suggested to him that the continents must once have been joined, and must, over time, have wandered

'If we are to believe Wegener's hypothesis we must forget everything which has been learned in the last seventy years and start all over again.'

R.T. CHAMBERLAIN, 1926

across the face of the globe, like huge icebergs slowly forcing their way through thinner pack ice. He retraced their wanderings to a point where they were all joined together in a super-continent he termed Pangaea (from the Greek for 'all land'). What force might conceivably drive such epic migrations he could not say for sure. Perhaps foolishly, however, he was willing to speculate that a *Pohlfluct* ('flight from the Poles') and some form of tidal friction might be jointly responsible.

The gathering storm

Right from the start, Wegener faced criticism for his bold attempt to cut across disciplines and forge a radical new theory. His father-in-law, who was a respected meteorologist, tried to dissuade him as early as 1911. Wegener defended himself: 'I believe that you consider my primordial continent to be a figment of my imagination, but it is only a question of interpretation of observations ... Why should we delay in throwing the old concept overboard? Is this revolutionary? I don't believe that the old ideas have more than a decade to live.' His optimism was ill-founded.

Criticism began soon after publication and continued for decades. In 1922, Philip Lake dismissed him as 'not seeking truth [but] advocating a cause ... blind to every fact and argument that tells against it.' Lake savaged the attempt to reconstruct Pangaea by fitting together the coastlines of the continents: 'It is easy to fit the pieces of a puzzle together if you distort their shape.' In fact, the true fit is between the continental shelves, but these were not well mapped at this point. A year later, G.W. Lamplugh described Wegener's theory as 'vulnerable in almost every statement', while R.D. Oldham wrote that 'it was more than any man who valued his reputation for scientific sanity ought to venture on.'

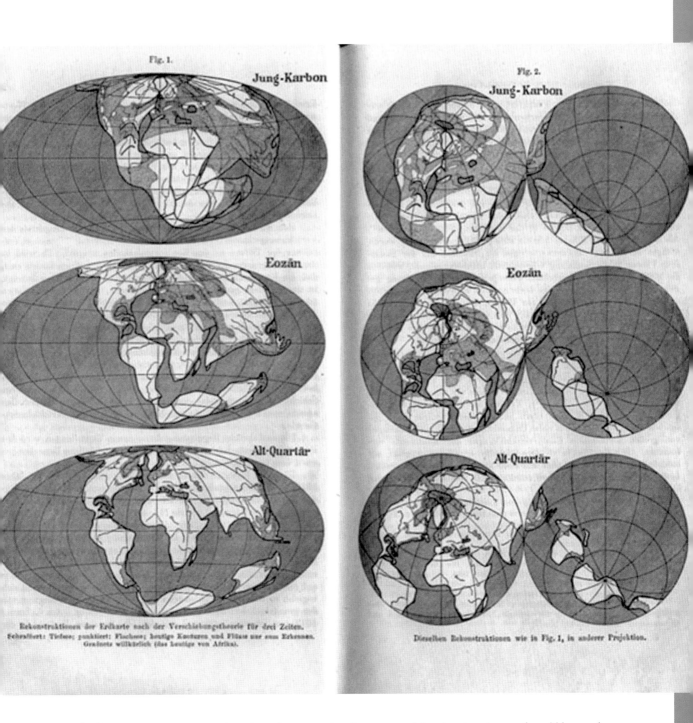

▲ Pangaea to present. A series of maps – showing equatorial and polar views – from Wegener's book, *The Origin of the Continents and Oceans*, showing the positions of the continents at different periods in geological history, beginning with his hypothetical supercontinent.

One of the most influential, trenchant and unrelenting critics of continental drift was the geophysicist Harold Jeffreys, who was particularly unimpressed with Wegener's notion of the continents sailing through a plastic ocean floor. He described it as 'a very dangerous [idea], and liable to lead to serious error.' Jeffreys calculated that the two forces Wegener had proposed as drivers of drift – a *Pohlfluct* and some form of tidal friction – could only provide one-millionth of the force that would be needed.

Opposition to continental drift came to a head at a symposium of the American Association of Petroleum Geologists held in New York in 1926, which both Wegener and Taylor attended. Attendees took turns to bash the theory. C.R. Longwell talked of, 'the very completeness of the iconoclasm, this rebellion against the established order … Its daring and spectacular character appeals to the imagination … But [it] must have a sounder basis than imaginative appeal.' Palaeontologist E.W. Berry accused Wegener of 'a state of auto-intoxication in which the subjective idea comes to be considered an

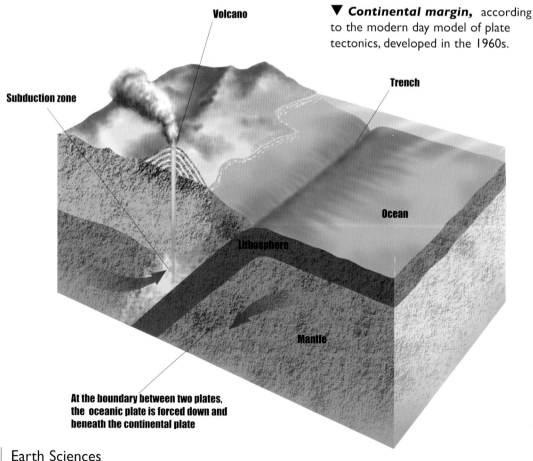

Volcano

▼ *Continental margin,* according to the modern day model of plate tectonics, developed in the 1960s.

Trench

Subduction zone

Ocean

Lithosphere

Mantle

At the boundary between two plates, the oceanic plate is forced down and beneath the continental plate

objective fact.' T.C. Chamberlain accused Wegener of 'taking considerable liberties with our globe', and his son, R.T. Chamberlain, would later wonder, 'Can we call geology a science when there exists such differences of opinion on fundamental matters as to make it possible for such a theory as this to run wild?' Harsh words were still being aimed at Wegener long after his death. In 1949, the revered geological engineer Bailey Willis described continental drift as 'a fairy tale'.

'And yet it moves'

Such vitriol, and the phrases deployed by his detractors, led many to compare Wegener to Galileo (*see* pages 166–171). Indeed, *Our Mobile Earth*, a book by a supporter of continental drift, Reginald Daly, included as an epigram Galileo's famous alleged observation '*E pur si Muove*' – 'And yet it moves'. Wegener's supporters felt vindicated when advances in oceanography and geology after the Second World War seemed to prove he had been correct after all. Bands of magnetic anomalies in ocean-floor rocks showed that the rocks had indeed been spreading, and the discovery of a network of mid-oceanic ridges where volcanic activity was pumping out new seabed explained why.

In 1960, Henry Hess proposed the spreading sea-floor hypothesis, explaining, 'The continents do not plough through oceanic crust impelled by unknown forces, rather they ride passively on mantle material as it comes to the surface at the crest of the ridge and then moves laterally away from it.' By 1965, Tuzo Wilson had synthesized the new discoveries into a comprehensive theory of plate tectonics, explaining how and why continents drifted, oceans spread, mountains were created, rift valleys opened, volcanoes erupted and islands formed.

So, was Wegener vindicated, like a latterday Galileo or Darwin? His overarching contention was proved correct – the continents had indeed come together to form super-continents in the distant past, and then had drifted apart to their current positions, and this did explain the distribution of mountain ranges, fossil and coal beds and much else besides. Yet Wegener had also been proved hopelessly wrong in many of the details of his theory, just as his critics contended. Although the plate tectonics model includes a form of continental drift theory, it is by no means the same as Wegener's version. The story of science is always more complicated than one man triumphing over others.

WILLIAMS
vs
CHOUET

FEUDING PARTIES
Stanley Williams (born 1952)
– volcanologist, professor of
geology
vs
Bernard Chouet (born 1945)
– volcanologist with the US
Geological Survey

DATE
1993

CAUSE OF FEUD
Whether the best predictors of an
imminent eruption – specifically,
the eruption of Galeras volcano
in Colombia – are seismological
events or levels of gas emissions

The protagonists of this 'feud' never really faced off in person or in the arenas of scientific debate. In fact, it is probably a little unfair to describe their relationship as a feud, although Williams is on record as saying that he and Chouet 'never really got along'. The two men are volcano experts, men whose driving ambition has been to work out how to predict an eruption. They differed sharply on their answers to this question, differences reflected in the ways they have approached their topic and in their views of what constitutes a 'real' volcanologist. These differences would be brutally and tragically highlighted in 1993, when Williams led a group of scientists and others into a volcanic crater, only for it to erupt, killing nine and severely injuring Williams himself.

Desk jockeys vs volcano jocks

Arguably, the world of volcanologists can be divided into jocks and nerds. The nerds work in laboratories and behind computers, poring over graphs, statistics and computer models. The jocks climb active volcanoes and get up close and personal with them. Williams has claimed that 'the best work … comes from those of us who walk into the crater', so it is probably fair to describe him as a volcano jock. In the words of Dr Larry Malinconico of Lafayette College in Pennsylvania, 'Stan's a very aggressive scientist.'

Williams undertakes dangerous ventures into the volcano's mouth, partly in order to study and collect samples of the noxious gases issuing from vents known as fumaroles. He believes that the make-up and volume of these emissions can be used to determine how close to an eruption the volcano is. Bernard Chouet, who trained as an engineer and rocket scientist before becoming a volcanologist, concentrates more on studying the seismology of volcanoes; he can be classed as one of the nerds. It was while studying seismographs in the early 1980s that he identified a hitherto unrecognized component of the vibrations produced by pre-eruption volcanoes: 'It stared you in the face. "Wow, this is obviously different." Embedded in the record among all these … earthquakes were classic-looking quasi-monochromatic harmonic signatures, beautiful textbook examples.'

The 'harmonic signatures' were what is more commonly known as long-period events (LPEs), known in Spanish as *tornillos* (screws) because of their characteristic corkscrew shape. They are resonant frequencies produced by bottled-up lava and gas vibrating inside volcanic fissures like air in an organ pipe, and Chouet realized that their presence signalled an imminent eruption. He described his revelation as 'a defining moment … suddenly you realize the volcano is speaking to you and you understand the language.'

▶ **Bernard Chouet.** Swiss geophysicist and seismologist who originally studied astronautics, making him, literally, a rocket scientist. Chouet is now best known for his pioneering research on Long Period Events.

Under the volcano

The pressing need for a reliable way to predict an eruption was driven home in 1985, when the Colombian volcano Nevado del Ruiz became highly active. Local volcanologists feared that when it erupted it would trigger catastrophic floods and mudslides that could threaten the nearby town of Armero, but without precise knowledge of when such an event might occur the authorities would not agree to evacuate over 20,000 people for an indefinite period. On 13 November the mountain blew, drowning Armero in a vast tide of mud, water and rubble that claimed 24,000 lives. Williams and Chouet continued work on their respective theories, hoping to prevent the next catastrophe.

In 1989, Chouet's LPE model proved itself when he was able to issue warnings to evacuate workers from an oil facility near the Alaskan volcano Mount Redoubt just two hours before it erupted. News of this feat was slow to percolate through the world of volcanology, but soon another dangerous volcano in Colombia was to become a test for both men's theories. In 1991, Mount Galeras became active, drawing attention from around the world. In 1992, both Chouet and Williams were studying the volcano closely, and in July both men began to predict an eruption. Williams had noted increased rates of emission of sulphur dioxide and other gases, and Chouet and his team had noted the appearance of *tornillos* on their seismographs. On 16 July, a small eruption destroyed observation posts round the rim of the crater; both men felt vindicated.

In January 1993, Williams secured UN funding to host an international conference for volcanologists in Pasto, near Galeras. He managed to attract many distinguished experts from around the globe, although Chouet and his team were not able to attend. Williams, something of a showman, planned for the highlight of the conference to be a field trip into the crater, led by himself. He had checked emissions from fumaroles in the crater and was convinced there was no imminent danger of an eruption because sulphur dioxide levels were relatively low. But other volcanologists had begun to detect telltale *tornillos* since 23 December and questions were raised about the safety of the expedition.

On the eve of the expedition there was a meeting but accounts of the ensuing discussion differ. Colombian volcanologist Fernando Gil, who had previously worked with Chouet, was alarmed: 'We were concerned by these long-period events and what had happened when we'd seen them before.' But Williams and his colleagues, who were not seismologists, did not really understand what the problem was. Williams' associate John Stix admits, 'There was a concern, but we didn't really understand what those [long-period] events were telling us.' Williams claims that there was no consensus about the predictive value of LPEs: '… there was no such understanding [that *tornillos* might presage an eruption]. In the days before our trip into the crater, no one brought the tornillos to my attention or warned that the volcano might be poised to blow … Based on all available evidence, the consensus at the observatory was that Galeras was safe.'

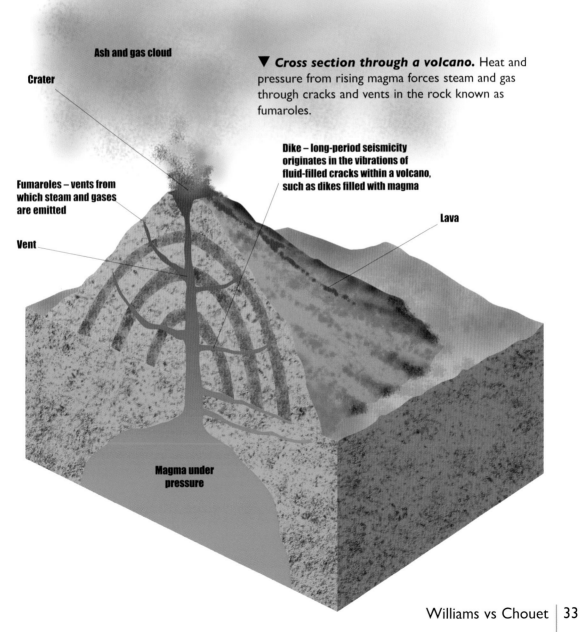

Ash and gas cloud

Crater

▼ Cross section through a volcano. Heat and pressure from rising magma forces steam and gas through cracks and vents in the rock known as fumaroles.

Dike – long-period seismicity originates in the vibrations of fluid-filled cracks within a volcano, such as dikes filled with magma

Fumaroles – vents from which steam and gases are emitted

Lava

Vent

Magma under pressure

TIMELINE

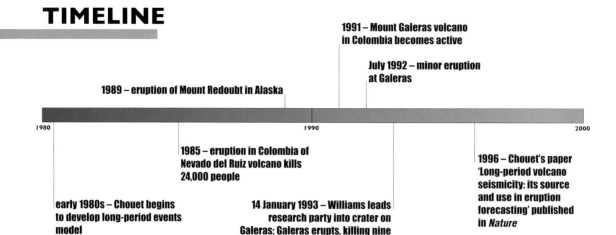

1991 – Mount Galeras volcano in Colombia becomes active

July 1992 – minor eruption at Galeras

1989 – eruption of Mount Redoubt in Alaska

1980   1990   2000

1985 – eruption in Colombia of Nevado del Ruiz volcano kills 24,000 people

1996 – Chouet's paper 'Long-period volcano seismicity: its source and use in eruption forecasting' published in *Nature*

early 1980s – Chouet begins to develop long-period events model

14 January 1993 – Williams leads research party into crater on Galeras; Galeras erupts, killing nine

Why was he not aware of the potential significance of Chouet's work? 'Unfortunately, Chouet – with whom I did not get along – never sent me a copy of his report,' he wrote in *Surviving Galeras*, his 2001 account of the disaster. Would it have made any difference? 'If I had been down there at the time and I had seen the long-period events I would certainly have voiced my opinion that it was not an appropriate time to go into the crater. But I couldn't have just jumped in front of them and said, "Over my dead body!" so I don't know what the outcome would have been,' points out Chouet.

Into the valley of death

The next morning, 14 January 1993, Williams led a party of nine scientists and three hikers up the mountain and into the caldera, the wide crater atop the volcano within which sits the volcanic cone itself. For several hours they took readings and gathered samples; other scientists watched from the caldera rim. At 1.30pm the volcano blew its top, blasting thousands of tons of rock into the air to rain down on those in and around the crater. Nine people were killed almost instantly, blown away or smashed by falling rocks. Williams himself was horribly injured, his leg mangled and slivers of skull driven into his brain.

Whether Williams was at fault for going ahead with the field trip is debatable but his behaviour after the disaster damned him in the eyes of many. He compounded bouts of apparent memory loss and fantasising with arrogance and contempt for others, in what science journalist Victoria Bruce, author of *No Apparent Danger: The True Story of Volcanic Disaster at Galeras and Nevado del Ruiz* and one of Williams' most trenchant critics, called 'a flagrant grab for fame at the expense of dead colleagues'.

▲ *Galeras erupting in 2008.* The fatal mountain continues to threaten surrounding communities, but better predictive science has improved the ability of the authorities to evacuate in time.

> *'I do not feel guilty about the death of my colleagues. There is no guilt. There was only an eruption.'*

STANLEY WILLIAMS, *SURVIVING GALERAS*, 2001

For instance, in a memo from the period Williams accuses two of the other survivors of being 'pathetic liars … jealous of the recognition which I received' for disputing his account. He now admits that he was 'playing the survivor', and some have ascribed his abysmal behaviour to the brain damage he sustained.

In the debate over which heralds of impending eruption – LPEs or elevated sulphur emissions – to heed, Chouet seems to have been vindicated. His 1996 paper on the use of LPEs for eruption forecasting is one of the ten most cited papers in volcanology. When the Mexican volcano Popocatépetl had its biggest eruption in a millennium on 18 December 2000, careful analysis of LPEs meant the authorities were able to give 48 hours notice to evacuate 30,000 people who lived in the danger zone. Not one person was hurt.

PART TWO
EVOLUTION and PALAEOBIOLOGY

HUXLEY
vs
WILBERFORCE

FEUDING PARTIES
T.H. Huxley (1825–95),
aka 'Darwin's Bulldog' – biologist,
educator
vs
Samuel Wilberforce (1805–1873),
aka 'Soapy Sam' – Bishop of
Oxford

DATE
Saturday, 30 June 1860, at the
meeting of the British Association
for the Advancement of Science
(BAAS) at Oxford

CAUSE OF FEUD
The descent of man

One of the most famous confrontations in science was the clash over evolution between T.H. Huxley, known as 'Darwin's Bulldog' because of his tenacity in defending Darwin's theory of evolution through natural selection, and Bishop Samuel Wilberforce, known as 'Soapy Sam' because of his smooth, persuasive, even slippery style in debates. This now legendary encounter is typically quoted as a milestone in the 'triumph' of science over religion, though many dispute this and there is a great deal of uncertainty over what was actually said.

Barking and yelping

As he knew it would, the 1859 publication of Darwin's *On the Origin of Species* aroused opposition from many quarters, including many in the Church. Evolutionary theories were common currency at the time but men like William Paley had reconciled them with religious belief by suggesting that evolution was directed by some sort of cosmic guiding hand (i.e. God). What was particularly shocking about Darwin's theory of evolution by natural selection was that it removed the need for a guiding intelligence directing proceedings. Blind forces could be responsible for all of life's many forms; there was no need for God in this conception of creation. Religionists were also uncomfortable with Darwin's unstated, though implied, conclusion that man might be descended from the apes, rather than being the result of a separate act of creation, made in God's own image.

TIMELINE

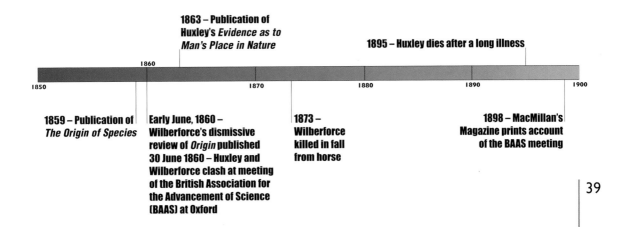

1863 – Publication of Huxley's *Evidence as to Man's Place in Nature*

1895 – Huxley dies after a long illness

1860

1850   1870   1880   1890   1900

1859 – Publication of *The Origin of Species*

Early June, 1860 – Wilberforce's dismissive review of *Origin* published
30 June 1860 – Huxley and Wilberforce clash at meeting of the British Association for the Advancement of Science (BAAS) at Oxford

1873 – Wilberforce killed in fall from horse

1898 – MacMillan's Magazine prints account of the BAAS meeting

39

Biologist T.H. Huxley immediately became an ardent acolyte of the new theory and, foreseeing that Darwin would face 'considerable abuse & misrepresentation', wrote to offer his services: 'as to the curs which will bark and yelp – you must recollect that some of your friends at any rate are endowed with an amount of combativeness which (though you have often & justly rebuked it) may stand you in good stead … I am sharpening up my claws and beak in readiness.'

OPENING PANDORA'S BOX: THE REACTION TO DARWIN

Darwin had been incubating his theory of evolution through natural selection for decades before he was finally pushed into publication by Alfred Russel Wallace's (see pages 186–191) independent arrival at a similar theory. One reason he had held off for so long was because he knew it would open a Pandora's box of reaction. In the event, the best he could do was marshal his arguments and evidence with minute care, and admit that there were a few gaps where he must trust to the future emergence of more evidence.

Wilberforce was one of the first to attack Darwin. His critique was powerful, pointing out flaws in Darwin's science – the lack of fossil evidence of transitional forms, for instance – and highlighting the disturbing social and moral implications of the theory. This line of attack is still pursued by religious critics of Darwinism today. In 2009, for instance, Christoph Schönborn, Cardinal Archbishop of Vienna, argued: 'The question of evolutionism and the economic crisis are very closely linked. What we can call the ideological Darwinist concept that the stronger survives has led to the economic situation we're in today.'

The religious response to Darwin's theory has been diverse and complex since it first emerged. The Catholic Church made little fuss about accepting the theory, partly because in Catholicism religious authority does not stem solely from scriptural sources. Many stripes of Protestant religion have also sought to accommodate Darwinism and there are many who argue that there is no contradiction between being a religious believer and a scientist, for instance. But there are also many at the extremes of religion, particularly in fundamentalist Christianity but increasingly in the Islamic world too, who loudly protest that any conception of evolution is at odds with religious belief, and it is these strident voices who have dragged Darwin into what is often referred to as the 'culture wars'.

▲ **'That man wants to claim my pedigree,'** complains the Defrauded Gorilla to animal rights campaigner Henry Bergh in a satirical cartoon of 1871. 'He says he is one of my descendants.' 'Now, Mr Darwin,' replies Mr Bergh, 'How could you insult him so?'

The barking and yelping soon began. Among Darwin's many detractors was Samuel Wilberforce, Bishop of Oxford. Son of the famous anti-slavery campaigner, Wilberforce was a high-profile public figure, a clergyman not afraid to weigh in to scientific debates. He published a dismissive review of *Origin* in June 1860, arguing that 'man's power of articulate speech; man's gift of reason; man's free will and responsibility … all are equally and utterly irreconcilable with the degrading notion of the brute origin [of humankind].' 'The principle of natural selection,' he concluded, was 'a dishonouring view of nature … absolutely incompatible with the word of God.'

'A man has no reason to be ashamed of having an ape for his grandfather. If there were an ancestor whom I should feel shame in recalling it would rather be a man who plunges into scientific questions with which he has no real acquaintance, only to obscure them by an aimless rhetoric.'

ONE VERSION OF T.H. HUXLEY'S LEGENDARY REJOINDER TO BISHOP WILBERFORCE, 30 JUNE 1860.

Monkey business

Just two weeks later, on Saturday, 30 June, Wilberforce was scheduled to give a paper at a meeting of the British Association for the Advancement of Science (BAAS) at Oxford. Here was a chance for Darwinists to challenge him in public. Darwin, stricken with the chronic illness that plagued him for life and reticent of public battles, declined to attend. Huxley was urged to go in his place, and so he took up the gauntlet. Hundreds flocked to witness the showdown, with many more turned away at the door.

Wilberforce, coached by the anti-Darwinian biologist Richard Owen (coiner of the term 'dinosaur'), delivered a fluent paper outlining many stinging criticisms of Darwin's theory. It was at the end of this speech, according to accounts, that he upped the stakes with a thinly veiled insult to Huxley. No precise transcript of the meeting exists, but one particularly popular account was that of Isabel Sidgwick, presented in *MacMillan's Magazine* nearly 40 years later: 'turning to his antagonist with a smiling insolence, he begged to know, was it through his grandfather or his grandmother that he claimed his descent from a monkey?' This was rough stuff for a Victorian audience, but the crowd was reportedly already raucous, with undergraduates chanting 'Monkey! Monkey!' provocatively.

HOPEFUL MONSTERS: THE PUNCTUATED EQUILIBRIUM DEBATE

There have been – and are – ongoing significant debates within the science of evolutionary biology. One of the most high profile has been the clash between the traditional view that evolution is a gradual, constant phenomenon, with species engaged in a slow but continuous process of evolution, and alternative models, of which 'punctuated equilibrium' is the best known. Briefly stated, this model suggests that speciation (the evolution of new species) occurs in rapid bursts lasting just a few millennia, making them almost invisible in geological terms, punctuating long periods when species are well adapted to their environments and change little if at all (i.e. in a state of 'equilibrium').

Huxley was among the first to propose a form of this theory, writing to Darwin just before the publication of *Origin* to warn: 'You have loaded yourself with an unnecessary difficulty in adopting *Natura non facit saltum* [Nature does not make leaps] so unreservedly.' In 1940, émigré biologist Richard Goldschmidt proposed in his book *The Material Basis of Evolution* that most mutants thrown up by evolution would not survive, but that once in a while an extreme mutation would occur that would cause a 'leap' in adaptive fitness. Perhaps unwisely he labelled these leaps 'hopeful monsters'.

Evolutionary biologist Stephen Jay Gould described how Goldschmidt's theory was regarded as 'anathema', while Goldschmidt himself 'became the whipping boy of [modern Darwinism]' . Together with the American palaeontologist Niles Eldredge, Gould revived the ideas of both Huxley and Goldschmidt in modified form, proposing the punctuated equilibrium model, partly in response to the troubling absence in the fossil record of the 'transitional forms' that had worried Darwin. Instead Gould had discovered fossil evidence of very rapid 'explosions' of speciation.

The debate between Gould and some of his opponents, most notably evolutionary biologist Richard Dawkins, has been portrayed as a latter-day continuation of the 19th-century geological debate between catastrophists and uniformitarians, but Dawkins himself argues that this is simply a crude misrepresentation of the arguments. He has dismissed Gould's theory as 'a minor gloss on Darwinism', which 'does not deserve a particularly large measure of publicity … the theory has been … oversold by some journalists.' Dawkins claims that in fact there is no opposition between punctuated equilibrium and Darwin's gradualism, and that 'The theory of punctuated equilibrium will come to be seen in proportion, as an interesting but minor wrinkle on the surface of neo-Darwinian theory.'

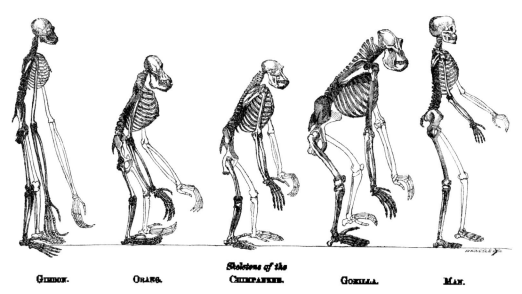

Skeletons of the

GIBBON. ORANG. CHIMPANZEE. GORILLA. MAN.

Photographically reduced from Diagrams of the natural size (except that of the Gibbon, which was twice as large as nature), drawn by Mr. Waterhouse Hawkins from specimens in the Museum of the Royal College of Surgeons.

▲ **Comparative anatomy.** To bolster his argument that humans and apes had a common ancestor, Huxley compared human skeletons to those of other apes, as shown here in the frontispiece from his book, *Evidence as to Man's Place in Nature*.

Sidgwick's account of his response was clearly intended to portray Huxley in a heroic light: 'On this Mr Huxley slowly and deliberately arose. A slight tall figure stern and pale, very quiet and very grave, he stood before us, and spoke those tremendous words – words which no one seems sure of now, nor I think, could remember just after they were spoken, for their meaning took away our breath, though it left us in no doubt as to what it was. He was not ashamed to have a monkey for his ancestor; but he would be ashamed to be connected with a man who used great gifts to obscure the truth. No one doubted his meaning and the effect was tremendous. One lady fainted and had to be carried out: I, for one, jumped out of my seat; and when in the evening we met at Dr Daubeney's, every one was eager to congratulate the hero of the day.'

Huxley's retort seems mild by today's standards, but the audience clearly inferred that he was saying he would rather be an ape than a bishop. Is this what really happened? According to Ronald Clark, a biographer of the Huxley dynasty, 'The details of what Huxley said differ as much as do those of Wilberforce's speech.' Huxley himself seems

INTELLIGENT DESIGN AND CREATIONISM

Creationists are those who believe in the truth of the biblical account of Creation. There are many different stripes of Creationist: Day-Age Creationists, for instance, believe that the six days of Creation mentioned in Genesis correspond to six ages or epochs of geological time; Young Earth Creationists believe that it took literally six days for God to create the Earth. Such an extreme belief is clearly at odds with Darwinian (or most other forms of) evolutionary theory, and fundamentalist Christians who advocate such beliefs try to challenge Darwinism on several fronts.

In America, where the Creationist lobby is powerful and vocal, there have been various attempts to limit the teaching of evolution in schools, such as the infamous Scopes Monkey Trial of 1925; or to ensure that some form of Creationism is taught alongside evolution. These attempts have mostly failed because the courts have deemed them to be attempts to teach religious beliefs in state-funded schools, which contravene the constitutional separation of Church and State in the United States.

In the late 1980s, a new movement arose propounding a theory known as Intelligent Design (ID), which is ostensibly a non-religious critique of Darwinian evolution that builds on the 'watch on the heath' argument proposed by William Paley in his 1802 book *Natural Theology*. Briefly stated, Paley's argument was that if you were out walking on a heath and came across a watch, you would assume that its intricate construction could not be the result of accident but that there must have been a watchmaker. This is a version of the teleological argument for the existence of God, which says that if there is some form of design or purpose to life/the universe, there must be a designer. Modern ID seeks to show gaps in Darwinism's account of the universe: for example, the eye is a complex organ that cannot have evolved through intermediate stages as it only confers a survival advantage in its 'final' form, in which all the components work together in perfect harmony. If it cannot have arisen by accident, the eye must therefore have arisen by design, which implies some form of intelligent designer.

The ID movement has attempted to circumvent America's Church–State separation rules by claiming that ID is a science and does not actually invoke God as the intelligent designer, that is is different from Creationism and should therefore be legal to teach in state-funded schools as an alternative explanation of the origin of life. Opponents of ID argue that this is transparent nonsense. In the 2005 Kitzmiller vs Dover Area School District trial, US Judge John E. Jones III ruled that ID is not a science and 'cannot uncouple itself from its creationist, and thus religious, antecedents'.

DAWKINS vs GOD

Richard Dawkins is an evolutionary biologist who first achieved fame with his books *The Selfish Gene* (1976) and *The Blind Watchmaker* (1986), which popularized modern Darwinian theory. He has become one of the figureheads of a new breed of assertive atheism, and his 2006 book *The God Delusion* has sold over 2 million copies. The book's assault on the dangerous and delusory nature of religion, together with his high-profile public pronouncements on the issues, have led former bishop Lord Harries to describe him as 'one of the attack dogs of fundamentalist atheism'. Harries, who recently debated with Dawkins at Oxford to mark the anniversary of Darwin's birth and commemorate the Wilberforce–Huxley clash, points out that 'The old atheism was content to say that Christianity was untrue. The new attack dogs also say it is dangerous … That's fighting talk.'

Dawkins pulls no punches in *The God Delusion*, describing the God of the Old Testament as 'arguably the most unpleasant character in all fiction'. He dismisses practitioners of what he calls 'understated, decent, revisionist religion' as being 'numerically negligible', focusing almost entirely on extremists, such as American television evangelists, or, as he puts it, 'crude rabble-rousing chancers'.

Dawkins' attack on religion has attracted a blizzard of criticism, most of it centred on what is perceived as his stridency and intolerance, which is often characterized as simply the mirror image of language used by religious fundamentalists. He dismisses criticism of his tone as pleas for exceptionalism on the part of religious belief: 'The illusion of intemperance flows from the unspoken convention that faith is uniquely privileged: off limits to attack.' More serious criticisms include that he swallows uncritically the myth of enlightened science at constant war with delusional and dishonest religion, that he ignores or is unaware of millennia of theological scholarship, and that by taking aim at only the crudest and most extreme caricature of religion he is prey to the 'straw man' fallacy (where an argument

◀ **Fighting talk.** *Spanish version of a global atheistic ad campaign, featuring posters on buses reading, 'There's probably no God. Now stop worrying and enjoy life.' The campaign has drawn fire from all sides – religionist and atheist.*

seems powerful because it fails to engage with the serious, substantive elements of the opposing view and concentrates only on easily disproved elements). Dawkins and others, such as American philosopher Daniel Dennett, who seek to apply the modern Darwinian philosophy to all areas of human experience, from psychology and spirituality to sociology and history, are also accused of unwarranted 'empire-building'.

Dawkins rebuts all these criticisms with varying degrees of success. For instance, he insists that 'Most believers echo Robertson, Falwell or Haggard, Osama bin Laden or Ayatollah Khomeini,' and points out, 'These are not straw men. The world needs to face them, and my book does so.' Journalist William Rees-Mogg, however, suggests that Dawkins' approach is fundamentally flawed, and fails to do justice to the legacy of Darwin: 'His tone is not like that of Charles Darwin himself; thoughtful, reflecting detailed observation, sensitive in the search for truth. It is more like that of Bishop Wilberforce in the Oxford debate of June 1860, in which the bishop attacked Darwinism.'

to have offered at least three different versions, and according to many accounts his immediate reaction to Wilberforce's sally was to mutter to a companion, 'The Lord hath delivered him into mine hands.' He knew that Soapy Sam had overstepped the mark, and went for the jugular in his response. Contemporary accounts of the meeting record nothing of either Wilberforce's 'grandmother' comment or Huxley's loaded response.

The legend of this encounter has grown over the years, whereas in reality it was probably little more than a minor skirmish in the protracted battles over evolution that continue to this day. For instance, Huxley's battle with Richard Owen continued. Owen claimed to have proved that man was not descended from the apes through his studies of brain anatomy, but Huxley marshalled proof to the contrary in his 1863 book *Evidence as to Man's Place in Nature*. Wilberforce went to an early grave in 1873, courtesy of head injuries sustained in a fall from a horse, which occasioned Huxley to remark uncharitably that 'reality and his brains came into contact and the result was fatal.' Today there are many who resent the supposed moral of the tale of Soapy Sam and the Bulldog – that Christians should stay out of science – which evangelical creationist Edward Coleson calls 'one of the most damaging pseudo-scientific myths to gain wide credence in the West in the last century or two'.

COPE

vs

MARSH

FEUDING PARTIES
Edward Drinker Cope (1840–97)
– palaeontologist
vs
Othniel Charles Marsh (1831–99)
– palaeontologist

DATE
1860s–90s

CAUSE OF FEUD
Priority over discovery of fossils
(aka the 'bone war')

The 'bone war' between Edward Cope and Othniel Marsh was the most notorious and damaging feud of its age, a Greek tragedy of hubris and nemesis, of two men locked together by obsession until one of them, goaded beyond endurance, unleashed the furies of public disgrace, bringing ruin upon both their heads. It was also a clash of personal ideologies that somehow embodied the late 19th-century battle for the soul of America, of individualism vs imperialism, libertarianism vs the establishment, anarchy vs order, a clash that would shape the settlement of the West.

A palaeontological education

Edward Cope and Othniel Marsh both came from money but their backgrounds and education were subtly different. Cope came from a genteel Quaker background and was schooled in the liberal tradition of a bygone age, touring the academic institutions of Europe during the Civil War. Marsh was older but began his schooling and career later, only beginning in earnest when his rich uncle George Peabody started to fund him; his education was more conventional and his career would later reflect this.

The two men had conflicting personalities and personalities suited to conflict. Marsh was not a naturally social creature; a college acquaintance observed that: 'for most people it was "like running against a pitchfork to get acquainted with him".' Meanwhile an acquaintance of Cope's, palaeontologist E.C. Case, wrote of him: 'he was essentially a fighting man, expressing his energy in encountering mental, rather than physical difficulties … He met honest opposition with a vigour honouring his foe, but fraternized cordially after the battle.'

The two men met for the first time in 1863. David Rains Wallace, author of the key text on the feud, *The Bonehunters' Revenge*, suggests that even then they probably 'felt a nascent rivalry … Their disparate backgrounds predisposed them to look down, subtly, on each other. The patrician Edward may have considered Marsh not quite a gentleman. The academic Othniel probably regarded Cope as not quite a professional.' Cope's freewheeling, individualist style contrasted with what Wallace calls the 'calm, methodical careerism' of Marsh, who quickly scaled the greasy pole of the scientific establishment.

'See the bones roll out'

Both men had conceived a passion for fossils, and were to become the greatest fossil hunters of their – and arguably any other – age. Awareness of fossils had gathered pace since the early 19th century and the end of the American Civil War, and the opening up of the American West, particularly the rampant growth of the railways, had set the stage for a blizzard of discoveries (*see* box, page 54). Cope and Marsh would be at the forefront of these developments. Scientific renown, even glory, was at stake for those who could find, reconstruct, describe and name new species, but what truly drove them was something dark and atavistic: an insatiable hunger to possess. Each man would eventually amass vast collections; in Marsh's case, at least, far more than he could adequately process.

At first there seems to have been a degree of cordiality, even cooperation, between the two men, but avarice soon overcame concord. Cope himself traced the beginning of their feud to 1868, claiming that he had taken Marsh on a tour of the New Jersey fossil beds, but that 'soon after, in endeavouring to obtain fossils from these localities, I found everything closed to me and pledged to Marsh for money considerations.' Marsh's financial clout (he was backed by his uncle's massive fortune, via the Peabody Museum at Yale) and political nous would increasingly enable him to 'reserve' fertile fossil sites as his private preserves.

Other reports date the start of the feud to an earlier incident in 1866, when Marsh published a report correcting an incorrect reconstruction of an elasmosaurus by Cope. The geologist Walter Wheeler, however, points to the summer of 1872, when both men were collecting in Bridger Basin, Wyoming; it appears it was at this time that their competitiveness boiled over into antagonism. The following year Marsh wrote to Cope to complain about his behaviour in Wyoming: 'The information I received … made me very angry, and … I was so mad … I should have "gone for you", not with pistols or fists, but in print … I was never so angry in my life.' Cope's response? 'All the specimens you obtained during August 1872 you owe to me.'

Their row grew to encompass arguments over access to fossils, accusations of deliberate destruction, attempts to hijack collections and bitter and complicated feuding over priority when it came to publishing descriptions of specimens. Marsh quickly retired from front-line collecting, hiring proxies to do the work (but claiming all the credit); Cope, without the funds or the institutional backing of his rival, was still in the field when the conflict reached its apogee at Como Bluff in Wyoming, one of the richest fossil sites in the world.

In 1877, each man was alerted to the site by different sources. Marsh's proxies resorted to code words and deception to throw Cope off the scent. His assistant Samuel Williston

TIMELINE

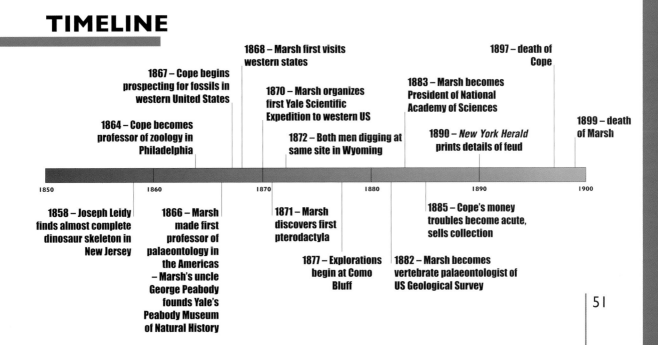

1868 – Marsh first visits western states

1897 – death of Cope

1867 – Cope begins prospecting for fossils in western United States

1870 – Marsh organizes first Yale Scientific Expedition to western US

1883 – Marsh becomes President of National Academy of Sciences

1864 – Cope becomes professor of zoology in Philadelphia

1872 – Both men digging at same site in Wyoming

1890 – *New York Herald* prints details of feud

1899 – death of Marsh

1850   1860   1870   1880   1890   1900

1858 – Joseph Leidy finds almost complete dinosaur skeleton in New Jersey

1866 – Marsh made first professor of palaeontology in the Americas – Marsh's uncle George Peabody founds Yale's Peabody Museum of Natural History

1871 – Marsh discovers first pterodactyla

1885 – Cope's money troubles become acute, sells collection

1877 – Explorations begin at Como Bluff

1882 – Marsh becomes vertebrate palaeontologist of US Geological Survey

▲ *Leg and foot bones* of a diplodocus, in situ, where they were discovered at Como Bluff. Remains of the diplodocus were first discovered here in 1878, and the genus was named by Marsh.

wrote to Marsh that the bones 'extend for seven miles and are by the ton.' Another assistant, William Harlow Reed, wrote to a colleague, capturing the excitement of the pursuit: 'I wish you were here to see the bones roll out and they are beauties … it would astonish you to see the holes we have dug.'

Cope was not to be denied, and soon rival camps were eyeing each other suspiciously across the site. There should have been plenty to go round, but stories abound of the depths to which the two men sank. Supposedly, Marsh ordered some dinosaur pits to be dynamited to prevent Cope from acquiring specimens, and later resorted to 'salting' Cope's excavations with random fossils that did not belong there, to obstruct his reconstructions. Cope allegedly responded by having a trainload of Marsh's findings diverted to Philadelphia. Tensions mounted and on at least one occasion guns were pulled. Wallace describes Como as 'more like one of the more squalid gold strikes than an instance of exalted scientific discovery.' He likens the antics of the two protagonists to industrial robber barons: 'In competing for a natural treasure – the abundant, unknown fossils of the western badlands – they might have been timber barons or mining tycoons.' Scientific ideals were now secondary; each man became as preoccupied with denying finds to his rival as in making them himself.

Dirty linen

While Cope was ploughing his own furrow out West, Marsh was back East becoming an establishment figure. The US Geological Survey (USGS) was a key player in the opening up of the West, and it could help or hinder access to fossil fields. Cope was associated with the Survey early on, but Marsh became a close ally of John Wesley Powell, its director, and soon Cope was ousted and Marsh installed as the official vertebrate palaeontologist. Marsh was winning the bone war, and as Cope's options and funds dwindled he became increasingly bitter. According to Wallace: 'He became Professor Moriarty to Othniel's academic Sherlock Holmes.'

Forced to sell most of his fossil collection in 1885, an increasingly desperate Cope was easily sucked into the orbit of one of the dark forces of the Gilded Age, newspaper magnate James Gordon Bennett, Jr. Wallace described Bennett as both 'a man as monstrous in his way as the palaeontologists' plesiosaurs and mosasaurs,' and at the same time, '[perhaps] the most underestimated American figure of the late nineteenth century.' In the pages of Bennett's mouthpiece, the *New York Herald*, the largest and most influential newspaper of the age, Cope opted to wash his dirty linen in the most public fashion imaginable. 'It is a business I do not like, but it is absolutely necessary,' he insisted.

Cope detailed a slew of charges of professional and personal misconduct against Marsh, and ensured that many of the Yale man's former associates were dragged into the feud (many were nursing grudges because of Marsh's high-handed, penny-pinching ways, and his insistence on sole authorship of papers deriving from fossil discoveries). Cope accused Marsh of restricting access to the fossil fields, abusing his role in the USGS through his relationship with Powell and deliberately blocking publication of Cope's findings. Marsh was guilty of plagiarism and taking credit for other men's work, and was not the true author of most of the papers attributed to his name. Bennett splashed

BONE RUSH

Fossils were the object of some confusion for much of history, and were commonly misinterpreted as the remains of mythical or magical beasts. Toadstones, for instance, were believed to be semi-precious stones sprung from the forehead or belly of a toad, which could warn of the presence of poison by growing hot. In practice, toadstones were probably fossilized fish teeth.

Palaeontology as a science traces its roots back to the 18th century and the comparative anatomical researches of French anatomist Georges Cuvier (1769–1832), who established that some fossil animals did not resemble living species, thus demonstrating the fact of extinction. The increasingly sophisticated study of fossils went hand in hand with developments in geology, and led to a growing appreciation of 'deep time' (geologic time scales). The epicentre of fossil hunting was in Europe, especially in Britain, but fossils had been discovered in North America as early as 1705, when a mastodon tooth was found on the banks of the Hudson River (it was initially identified by American clergyman Cotton Mather as the tooth of an antediluvian giant).

The golden age for American fossil hunting came with the opening of the West. After the Civil War there was a massive expansion of railroads, military bases, trade routes and settlement outposts across mid- and western America, and the excavation involved brought to light the unique geology of the region that makes it, in the words of palaeontologist Keith Parsons, 'a geologist's dream and a vertebrate paleontologist's paradise'. This region has been subjected to faulting, folding, uplift and erosion, exposing huge formations of sedimentary rock that contain some of the richest fossil sites in the world. Marine fossils, in particular, are in abundance because much of the mid-West had previously been submerged beneath a vast shallow sea (now known as the Western Interior Sea).

Notable sites include Como Bluff in Wyoming, where the conflict between Marsh and Cope came to a head. Fossils here were so richly abundant that a local trapper had built himself a cabin out of them, and Bone Cabin Quarry became one of the key fossil sites in the late 1890s. Another significant site is the Carnegie Quarry near Vernal, Utah. So many dinosaur bones were dug up from here that the collectors were exhausted before the quarry, and the site became Dinosaur National Monument, where fossils can now be viewed in situ. Many of the dinosaurs that take pride of place in the world's great natural history museums, which have done so much to transform museums into objects of childish enthusiasm, date back to the great 'bone rush' of the late 19th-century West.

> ## 'It is doubtful that any modern controversy among men of learning has generated more venom than this one did.'
>
> WALLACE STEGNER, *BEYOND THE HUNDREDTH MERIDIAN*, 1954

the details of this unedifying row across the front page of the *Herald* on the morning of 12 January 1890. The scientific community reeled.

But Cope had struck a Faustian pact; Marsh and his supporters hit back, with the result that both men's names were dragged through the mud. Marsh accused his detractors of being 'little men with big heads', prompting a former assistant Otto Meyer, to respond with a damning critique of his methods, concluding: 'I presume that all true scientists have more regard for a little man with a big head than for a big man with a little head.'

The *Herald* peddled the story for another two weeks, by the end of which the reputations of both men were irredeemably tarnished. It had lasting effects, dividing the world of American palaeontology into warring camps for decades. According to William Berryman Scott, professor of geology and palaeontology at Princeton from 1884 to 1930: '[This] most important feud … hindered and hampered the younger generation for years. Even yet, its effects persist … and crop out when one is least expecting them.'

According to Wallace, the differing ideologies of Cope and Marsh reflected America's own clash of ideologies: 'In a way, the bone war was an intellectual variant on the Civil War that just preceded it, one largely enacted in that other great arena of North American conflict, the West.' It also reflected two visions of the West, Cope's 'nativist', individualistic, libertarian approach contrasting with Marsh's urban, hierarchical, 'imperialist' bent. More concretely, the squabble had direct impact on how the West was won; the geological surveys in which Marsh and Cope were involved helped prepare the ground for eventual settlement, while the damage to Marsh's reputation tarred by association his ally Powell and derailed Powell's plans for limited, controlled colonization of the West and instead opened the way for the free-for-all that followed.

LEAKEY
vs
JOHANSON

FEUDING PARTIES
Richard Leakey (born 1944)
– British–Kenyan scion of Leakey
palaeontology dynasty and major
figure in African conservation
vs
Donald Johanson (born 1943)
– American palaeoanthropologist;
discoverer of 'Lucy'

DATE
1978 onwards

CAUSE OF FEUD
Dispute over identity of hominid
remains, complicated by media
profiles

The Leakey family are a remarkable dynasty of scientists and probably the most famous names in palaeoanthropology. Louis Leakey and his wife Mary brought to the world's attention the rich hominid fossil grounds of Olduvai Gorge in the East African Rift Valley, often described as the cradle of humanity and the original Eden. Their son Richard achieved an international profile thanks to his own sensational finds, his conservation work and his media work (books and television series). The Leakeys were the first to identify *Homo habilis*, the oldest member of the *Homo* genus dating back *c.*2.5 million years, and their efforts were concentrated on elucidating a *Homo* lineage stretching back still further, which would sideline the other major genus of human ancestors, *Australopithecus*, as a dead end in evolution that did not contribute to modern humans.

Lucy and the footprints

One of the most exciting Leakey finds was the 1978 discovery at Laetoli in Tanzania of a line of footprints in ancient volcanic ash. Made by at least two early hominids (with some suggestions that a third individual walked in the footprints of one of the other two), they provided incontrovertible proof of bipedalism dating back some 3.6 million years. For the Leakeys, this promised proof that the lineage they were uncovering stretched back even further than *Homo habilis*, and Mary was excited to have the chance to present her findings to a wider audience at a Nobel symposium in Sweden later that year, where she was to be honoured with a gold medal.

Unbeknown to the Leakeys, another palaeoanthropologist would also be discussing his findings at the symposium: Donald Johanson, the American who, four years earlier, had helped uncover the skeleton known as 'Lucy' (he named her for the song 'Lucy in the Sky with Diamonds', which was being played around camp). Lucy, a youngish female, had spectacular implications for palaeoanthropology, as she appeared to be a direct ancestor of the *Homo* genus, except that she was only around 3 million years old and so was too young. However, Johanson identified her as belonging to a new species of

> *'What if Leakey was right and Johanson was wrong? What difference would that make in the significant understanding of human evolution? You've got to say the answer is, none. It wouldn't change the pattern of evolution, just some of the details.'*

MILFORD WOLPOFF, PROFESSOR OF ANTHROPOLOGY, 1984

Australopithecus, which he named *afarensis* after the Afar region of Ethiopia, where she had been found. Extending her lineage back further would strengthen the importance of Johanson's new species, and the Laetoli footprints seemed to be exactly the evidence he needed, despite being considerably older and more than a thousand miles distant from Afar.

Johanson spoke first, and talked at length about the Laetoli footprints, declaring that they had been made by none other than *Australopithecus afarensis*. According to reports, Mary was dismayed. 'How am I going to give my paper now? It's all been said,' she complained to Richard, who later explained that she felt she 'was going to look as if she was a fool, repeating the material'. Mary was particularly put out that Johanson had appropriated the footprints as evidence for his favoured genus, rather than *Homo*, expressing regret that 'the Laetoli fellow is now doomed to be called *Australopithecus afarensis*.'

The next generation

Battlelines now seemed drawn and, as the public face of the Leakey dynasty, it fell to Richard to take up arms against Johanson. At least this was the narrative according to

▶ **Mary Leakey** pictured at work on the famous Laetoli footprints in Tanzania. The structure of the prints provides clear evidence that whatever made them had an upright gait similar to that of modern humans.

HUMAN EVOLUTION TIMELINE

1.6 mya – Migration out of Africa of *Homo erectus*

3.3–2.5 mya – Evolution of *Australopithecus africanus*

c. 200 kya – Evolution of *Homo sapiens*

4mya   4mya   2mya   1mya

100kya

4–3 mya – Evolution of *Australopithecus afarensis*

2.5–1.9 mya – Evolution of *Homo habilis*

1.9 mya – Evolution of *Homo ergaster* (African lineage of *Homo erectus*)

c. 90 kya – Migration out of Africa of *Homo sapiens*

c. 50 kya – First *Homo sapiens* in Australia

c. 40 kya – First *Homo sapiens* in America by coastal route?

c. 11.5 kya – Clovis peoples arrive in America by land route

c. 12 kya – *Homo floresiensis* still living in Indonesia?

mya = million years ago
kya = thousand years ago

the media, which portrayed the row as a fight between the old guard, in the form of the Leakeys and their eponymous foundation, and the upstart Johanson, who founded his own organization, the Institute of Human Origins, and who was steadily building a profile to match Richard's.

The two were pitched head to head in a 1981 episode of the US television programme 'Cronkite's Universe', hosted by Walter Cronkite. The veteran newsreader set up the encounter provocatively, explaining, 'Before the discovery of Lucy made Donald

◀ **Louis and Mary Leakey,** studying some of their finds. Their prolific careers spanned many decades and radically enhanced our knowledge of the human family tree.

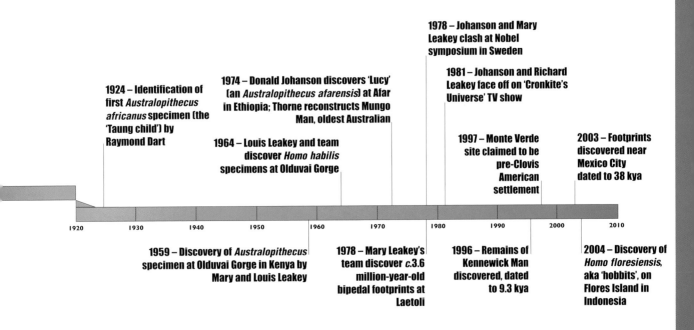

1924 – Identification of first *Australopithecus africanus* specimen (the 'Taung child') by Raymond Dart

1974 – Donald Johanson discovers 'Lucy' (an *Australopithecus afarensis*) at Afar in Ethiopia; Thorne reconstructs Mungo Man, oldest Australian

1964 – Louis Leakey and team discover *Homo habilis* specimens at Olduvai Gorge

1978 – Johanson and Mary Leakey clash at Nobel symposium in Sweden

1981 – Johanson and Richard Leakey face off on 'Cronkite's Universe' TV show

1997 – Monte Verde site claimed to be pre-Clovis American settlement

2003 – Footprints discovered near Mexico City dated to 38 kya

1959 – Discovery of *Australopithecus* specimen at Olduvai Gorge in Kenya by Mary and Louis Leakey

1978 – Mary Leakey's team discover *c.*3.6 million-year-old bipedal footprints at Laetoli

1996 – Remains of Kennewick Man discovered, dated to 9.3 kya

2004 – Discovery of *Homo floresiensis*, aka 'hobbits', on Flores Island in Indonesia

1920 1930 1940 1950 1960 1970 1980 1990 2000 2010

Johanson a celebrity, the king of the mountain of paleoanthropology was Richard Leakey.' A visibly unsettled Leakey was confronted by a chart of the hominid family tree according to his rival, and drew a cross through it, writing in its place a question mark. 'I'm not going to say whether you are right or wrong, Don,' he commented, 'but I think you're wrong.'

The dispute has rumbled on since then, mostly respectfully, but not always. A colleague of Johanson, Tim White, who had previously worked with the Leakeys, claimed that Mary's crew had altered the Laetoli prints with a tool during the excavation, a charge Mary dismisses as 'nonsense'. When Richard made his most famous find, the 1984 discovery of a complete *Homo erectus* skeleton nicknamed the 'Turkana Boy', Johanson commented snidely, 'I'm a little surprised that everyone is so surprised.'

In her autobiography, *Disclosing the Past*, Mary Leakey claimed that new findings bringing forward the date of Lucy showed that Johanson's fossils 'are too young to occupy an ancestral position, even if they were otherwise appropriate'. Johanson countered by accusing the Leakeys of making things personal: 'I think they feel a lot of what we've done has been a personal attack on their family. The most frustrating aspect of it is that neither one of them, Richard or Mary, will sit down and discuss their criticisms in a scientific journal. They have chosen to criticize us in the press.' Nevertheless, Johanson has also spoken of his respect for all that the Leakeys have achieved, specifically for 'finding many specimens of great importance'. The primary response of the Leakeys has been to withdraw from the debate, let their findings do the talking and, in the case of Richard, largely retire from fossil hunting to concentrate on conservation.

PALEOANTHROPOLOGY: THE MOST CONTENTIOUS SCIENCE

Controversy thrives in a vacuum, and palaeoanthropology, the science of human origins, or, in the parlance of the popular press, the hunt for the missing link, is plagued with gaps. From the skulls, skeletons, tools and footprints of a few dozen individuals, scientists have attempted to trace the course of human evolution and build up a family tree of hominids (the family that includes the genus *Homo* and the species *Homo sapiens* – modern humans), but because the evidence is scant and decent hominid fossils are so rare, a single skull or even tooth must sometimes bear a huge weight of interpretation. Scientists build their careers and reputations around particular interpretations, so that academic disputes easily spill over into professional and personal ones.

One of the key divisions is between 'splitters' and 'lumpers' – those who use the evidence to split the human lineage into multiple species, and those who choose to group remains into fewer species, interpreting the differences as the natural variation expected within species. For example, according to 'lumpers', *Kenyanthropus* and *Paranthropus* (two genera recognized by 'splitters') are simply species of *Australopithecus*. As Milford Wolpoff, who falls into this group, points out: 'You see more variations in our own species on any street corner than you can see in some of these fossils.'

Out of Africa vs Multiregionalism

The greatest debate in the study of human origins is between two competing hypotheses about how modern human populations arose. One hypothesis, known as multiregionalism, holds that modern humans have evolved from a very diverse, widespread species of human that initially evolved in Africa but migrated out of the continent over a million years ago, rapidly spreading across Eurasia. Local populations adapted to local conditions, but there was sufficient interbreeding between these populations to ensure that they remained essentially a single species, even as they evolved from the early, ape-like form of *Homo erectus* through archaic forms of *Homo sapiens* into modern *Homo sapiens*. This theory explains why there seems to be continuity between features found in fossils and modern populations in the same area (although opponents of this model, such as Allan Wilson and Rebecca Cann, argue that, 'The continuity seen by believers in multiregional evolution may be an illusion') and is based on the assumption that different populations of humans, some more archaic than others, would have interbred – for instance, Neanderthal man and Cro-Magnon Man. 'Sex happens. I find this neither disturbing nor surprising,' comments Erik Trinkhaus, alongside Milford Wolpoff and Alan Thorne, a prominent proponent of the multiregional hypothesis.

The Out of Africa (OOA) hypothesis – also known as the Recent African Origin theory, the Recent Single Origin hypothesis or the Replacement model – holds that all modern humans are descended from a small group of

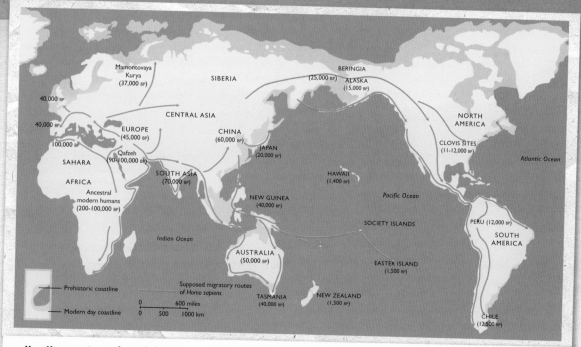

How Homo sapiens colonized the globe, according to the Out of Africa hypothesis. Note that almost every date on the map is contentious. BP = Before Present.

already anatomically modern *Homo sapiens* who left Africa roughly 90,000 years ago at most, having evolved there *c.*200,000 years ago. These recent migrants spread rapidly around the globe, replacing (and possibly wiping out) all other species of *Homo*, such as descendants of the much earlier *Homo erectus* global colonization. Interbreeding between the groups was minimal or non-existent and played no part in producing the genetic make-up of modern humans.

> **'Practically all palaeontological discoveries can be described as bones of contention'**
>
> John Napier, anthroplogist, The Roots of Mankind, 1971

The OOA theory is accepted by the majority of palaeoanthropologists (including major names, such as Chris Stringer and Stephen Oppenheimer) and is widely regarded as the mainstream view, backed up by the weight of fossil evidence and, especially, genetic evidence. By analysing the presence and evolution of certain marker gene variations, it is possible to deduce when and where these variations arose and, therefore, when and where different populations diverged. Although there is widespread variation within Africa, dating back 100,000 years or more, variations found outside Africa are limited and can all be traced back to a small group, perhaps even to a single individual, who lived around 80,000 to 70,000 years ago. This finding strongly suggests that shortly before or after this time these people/person left Africa to begin the colonization of the globe. Recent comparative study of African and Asian artefacts from around this period

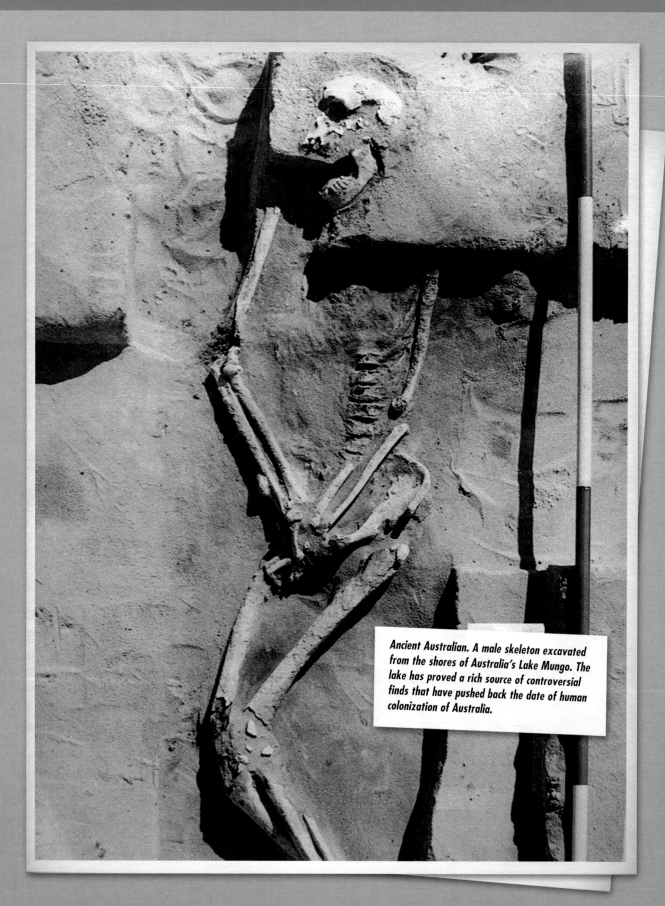

Ancient Australian. A male skeleton excavated from the shores of Australia's Lake Mungo. The lake has proved a rich source of controversial finds that have pushed back the date of human colonization of Australia.

further backs up the OOA hypothesis, although multiregionalists claim there is significant evidence to the contrary, such as the bones of Mungo Man.

Mungo Man

Lake Mungo in Australia is a rich source of prehistoric human remains, including the skeleton of an individual called 'Mungo Man'. Alan Thorne, who helped to discover and piece together the skeleton, argues that Mungo Man disproves the OOA hypothesis because, despite being anatomically modern and sprinkled with red ochre, evidence of a burial ritual similar to that still practised by Aboriginal inhabitants of the region, he was more than 60,000 years old. Thorne claims that this is far too old to fit the OOA model, in which *Homo sapiens* first left Africa not much more than 10,000 to 20,000 years before this date. His case appeared to be bolstered by an analysis published in 2001 in which, amazingly, DNA was recovered from the bones and analysed, and proved to contain a sequence not found in modern humans. According to Alan Mann, an anthropologist at Princeton University, 'The people at Mungo were totally modern looking and were expected to carry the DNA we have, but they didn't. I think that makes for an incredibly complicated story. It's a stunning development.' For Thorne this was conclusive evidence; speaking in 2002, he cited Sherlock Holmes' dictum about ruling out the impossible (so that whatever is left, no matter how improbable, must be true): 'Out of Africa is the impossible, and regional continuity is not only not improbable but the answer and the truth.'

Since then, proponents of the OOA model have poked holes in Thorne's story. Critics claim that the recovery of DNA from Mungo Man was suspiciously successful (i.e. despite safeguards, contamination could be involved), and point out that the unusual DNA sequence identified could be one that simply died out with Mungo Man's lineage, which would have no bearing on the validity of the OOA model. More recently, the man who actually found Mungo Man, James Bowler, has claimed that Thorne's original dating was misleading because it was based on sand taken from too far away from the burial site. Using sand from the site itself, Bowler has arrived at a new date of 40,000 years, and argues that the earliest human occupation around the Lake was no more than 50,000 years ago. As Chris Stringer of the Natural History Museum in London points out, the original human migrants would have had to travel less than a mile a year to get from Africa to Australia in the time available, making the journey eminently feasible.

KETTLEWELL

vs

HOOPER AND WELLS

FEUDING PARTIES
Bernard Kettlewell (1907–79)
– ecological geneticist and
entomologist
vs
Judith Hooper (born 1949)
– journalist and author of
Of Moths and Men;
Jonathan Wells (born 1942)
– fellow of the Discovery
Institute

DATE
1998 onwards

CAUSE OF FEUD
The peppered moth, icon of
evolution

Spectators might feel that a feud between a dead man and a number of living antagonists is one-sided at best, but Bernard Kettlewell has a legion of able supporters among the living. Since a furore broke out in 1998 over his landmark work on industrial melanism in the peppered moth, evolutionary biologists have rallied round to defend his good name and accuse his detractors of sour grapes, shoddy journalism and intellectual dishonesty.

Moth man

Bernard Kettlewell was a colourful and controversial character. He had trained as a doctor but gave up medicine and moved to South Africa after the Second World War to pursue his passion, entomology. In Africa he travelled across the continent collecting insects, especially moths, and carrying out field experiments that were 'admirable for their ingenuity and requirements for diligence', according to his biographer Geoffrey Morson. Morson relates that 'his personality … was outgoing and full-blooded; this led him to engage in sometimes maddening but good-natured controversy with his peers' and writes: 'Like, that other great truant from medicine, Charles Darwin, Kettlewell's results, as well as his methods, illustrate the related propositions that there need not be a barrier between the gifted amateur and the specialized academic and that important and lasting science is done by those who can combine the best elements of those seemingly opposite approaches.'

Not everyone has taken such a rosy view of Kettlewell's eccentric blend of amateurism and academia, which found its fullest expression when he returned to England to take up a post at Oxford under geneticist Edmund Brisco Ford (1901–88). In a series of experiments in the mid-1950s, Kettlewell sought to find what he called 'Darwin's missing evidence' – a clear illustration of natural selection in action. He succeeded through ingenious study of a phenomenon known since Victorian times, when it had been noted that the peppered moth, *Biston betularia*, appeared to be changing colour from mottled pale to dark (melanistic). It was widely speculated that this might be due to industrial pollution killing the pale lichen and blackening the bark of trees where the moth would rest, making it easy for predatory birds to spot pale moths, while giving a survival advantage to those few melanistic moths that existed thanks to natural variation.

Kettlewell undertook painstaking experiments to capture, mark and release different types of moth in woodland tainted by pollution to different degrees, then recapture as many as possible the next day to check relative survival rates. He and assistants also observed bird predation of moths. This work seemed to prove that there was indeed a survival advantage to melanism in polluted woodlands, and that this could account for the change in frequency of melanistic moths. Kettlewell's photographs of pale and melanistic moths against tree trunks with polluted and unpolluted bark quickly became fixtures in every standard textbook of biology, and he won renown and scientific glory for his experiments. The peppered moth was 'the slam-dunk of natural selection' and quickly became 'evolution's number one icon' according to Judith Hooper. Kettlewell himself came to a sad end, dying from an overdose of pain medication.

The truth about Santa

In 1998, Cambridge Professor of Ecology Michael Majerus (1954–2009) published a new book, *Melanism*, which made some criticisms of Kettlewell's work and pointed out some minor flaws in his conclusions. Unwittingly, Majerus had opened a can of worms. Ted Sargent, an American biologist who had long opposed Kettlewell's conclusions, was emboldened to write a scathing critique of his own, arguing that 'there is little persuasive evidence, in the form of rigorous and replicated observations and experiments to support [Kettlewell's] explanation [i.e. bird predation as the agent of selection].' Another biologist, Jerry Coyne, in a now infamous review of Majerus' book in the journal *Nature*, wrote

◀ **Bernard Kettlewell,** the eccentric British entomologist and geneticist, collecting insects in Brazil in the late 1950s. By this time he had already begun the research that was to make him famous, back in Britain.

> *'Hooper and Sargent should have performed a careful analysis before Hooper presumptuously insinuated fraud ... Hooper's claims are moonshine.'*
>
> MATT YOUNG, PHYSICIST, 2004

emotively: 'My own reaction [to Majerus's revelations about Kettlewell] resembles the dismay attending my discovery ... that it was my father and not Santa who brought the presents on Christmas Eve.' He reached the dramatic conclusion: 'for the time being we must discard *Biston* as a well-understood example of natural selection in action, although it is clearly a case of evolution.'

These rumblings reached the ears of science journalist Judith Hooper, who decided to take a closer look. What she uncovered, or thought she had uncovered, was startling, with great repercussions for evolutionary science and its ceaseless battle with creationists. In her acclaimed book *Of Moths and Men*, Hooper related a quite different version of Kettlewell's partnership with Ford. According to Hooper's version, Kettlewell had been an innocent abroad in the intellectual jungle of Oxford, driven to distraction by Ford, a manipulative schemer who had used his insecurities against him and cruelly blocked his attempts to gain acceptance to the Royal Society. In an attempt to appease his tormentor, Kettlewell had been driven to modify, perhaps even to manipulate his results.

The smoking gun in Hooper's case was a letter Ford had written in 1953, variously described in conflicting accounts of the research as 'encouraging', 'demanding' or 'critical', on the subject of Kettlewell's low rate of success in his crucial release–and–recapture experiments. Immediately after receiving this letter, Kettlewell's recapture rates rocketed, providing the basis for his landmark conclusions. Having failed to find any obvious meteorological explanation for the sudden improvement, Hooper asks, 'is it possible that [Kettlewell] made modifications in his experimental design?' A few pages later, she writes, 'what had passed unnoticed by their peers for at least a decade, was that Bernard had done a little tweaking ... in Birmingham in 1953.' Eventually, she claims: 'The unspoken possibility of fraud hangs in the air.'

Hooper points to several other problems with the moth experiments, and quotes Ted Sargent to damning effect on Kettlewell's scientific shortcomings – 'We don't allow experiments like this any more' – and apparent self-delusion – 'There are subtle ways to seduce yourself.' She argues that the whole episode is a cautionary tale of 'flawed science, dubious methodology and wishful thinking', and concludes, 'Clustered around the peppered moth is a swarm of human ambitions, and self-delusions shared among some of the most renowned evolutionary biologists of our era.' Perhaps most shockingly she even claims that Kettlewell was effectively driven to suicide by Ford (although most people argue that his overdose was probably linked to chronic pain he suffered after falling out of a tree during one of his field experiments).

Peppered myth?

Revelations such as those contained in Hooper's book have been manna for creationists and proponents of intelligent design (ID) in their crusade against evolution (*see* page 45). Leading the attack on what he calls 'the peppered myth' is Jonathan Wells, a senior fellow at the Discovery Institute, founded to promote the ID cause. Wells' 2000 book, *Icons of Evolution*, presents the case against the peppered moth (together with several other 'icons'), touting evidence such as the fact that Kettlewell's famous photos were staged by pinning dead moths to logs, and the finding that moths relatively rarely actually rest on tree trunks.

Wells' 'evidence' has aroused the ire of the evolutionary science community; among many who have taken issue with him are Kevin Padian and Alan Gishlick of the National Center for Science Education, who say, 'Wells likes to imply that there is a vast conspiracy of scientists protecting some alleged myth about the importance of selection in the case of the peppered moth.' Wells has accused Padian of 'Character assassination supported by transparently bogus statistics', and of comparing him to the sociopathic protagonist of the film *The Talented Mr Ripley*. 'We did no such thing,' insist Padian and Gishlick, 'Wells is likening himself to a sociopath. That's his privilege, but he is putting his words in other people's mouths.' However what they do imply is that Wells, like Ripley, is: 'merely an opportunist whose spiraling lies and envy trap him into increasingly desperate acts'.

This spat is simply a sideshow to the main event, which has been a wholesale refutation of the attacks on Kettlewell and the peppered moth experiments. Majerus, for instance, contends: 'The suggestion that Kettlewell ever "faked" a result is offensive to his memory. He was an honourable, good scientist who reported his findings with honesty and integrity.' Far from being disproved, his findings on industrial melanism have been robustly confirmed by multiple studies replicating and extending his work. According to Jim Mallet of the prestigious Galton Laboratory, 'as [Laurence] Cook [in a paper published in 2000] has demonstrated, even if Kettlewell was a fraud (and there is no good evidence that this was the case), the other 30-odd experiments on survival of adult moths in the field done by different scientists are convincing on their own.' Mallet concludes: 'the peppered moth story is about as convincing an example of natural selection by bird predators as you could possibly hope to find.'

▲ *Industrial melanism.* The peppered moth at the bottom is melanistic, camouflaged against bark stained with soot. Against unpolluted, lichen-covered bark, the lighter-coloured moth (top) is better camouflaged and more likely to avoid

What of the specific objections to Kettlewell's work? It is true that the famous photos were staged, but this was never denied or 'covered up'. It is also true that modern findings show that peppered moths only rest on tree trunks about 25 per cent of the time; the rest of the time they rest on the underside of branches, at branch-trunk junctions and on leaves – all of which are affected by industrial pollution and none of which invalidate the basic findings. Even the suspicious letter from Ford that supposedly prompted Kettlewell to engage in fraud is not what it seems, for it transpires that it was written the day *after* recapture rates had started increasing.

The aspersions cast on Kettlewell's work have led many textbooks to drop the famous photos and leave the peppered moth out of the story of evolution altogether. According to biology professor Bruce S. Grant this is a big mistake: 'The case for natural selection in the evolution of melanism in peppered moths is actually much stronger today than it was during Kettlewell's time. Textbook accounts should be expanded to reflect this newer information, and they should not cite *Of Moths and Men* as a credible resource.'

PILTDOWN MAN AND THE PREHISTORIC CRICKET BAT

Despite the protestations of the Intelligent Design movement (*see page 45*) and the suspicions of Judith Hooper, Kettlewell's peppered moths were probably not a scientific fraud intended to bolster the case for evolution by manufacturing 'Darwin's missing evidence'. In 1910, however, a fraud was perpetrated that fitted this profile exactly: the Piltdown Man hoax.

The late 19th and early 20th centuries was a period of popular fascination with the so-called 'missing link' – putative half-man, half-ape ancestors that would 'prove' Darwinian theories of the descent of man. Against this background, Charles Dawson (1864–1916), an amateur fossil hunter with strong links to professional palaeontology – notably his friendship with Arthur Smith Woodward (1864–1944), keeper of geology at the British Museum (now the Natural History Museum) – made a remarkable discovery. According to him, in around 1908 he was presented with a piece of human skull by a workman at the gravel pit at the village of Piltdown in Sussex (southern England). Excavating for himself, he turned up more pieces of skull, and in 1912 he called in his friend Woodward to help. Together they excavated more of the Piltdown site, uncovering a treasure trove of prehistoric finds, including stone tools, fossil animal bones and a jaw to go with the skull.

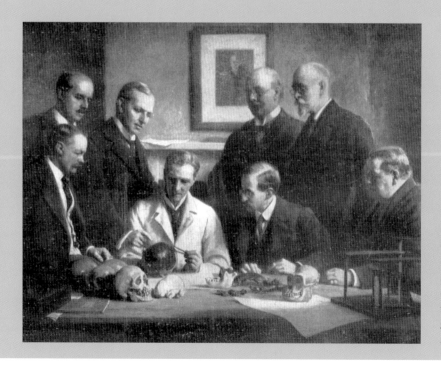

◀ **Luminaries of British anthropology** *examine the skull of Piltdown Man. Dawson and Woodward are the two gentlemen standing on the right of the picture.*

▶ **Reconstruction** *of the face of Piltdown Man, from the* Illustrated London News *of 28 December 1912.*

In December 1912, Dawson presented his discoveries to a meeting of the Geological Society of London. The Piltdown skull and jawbone caused a sensation as they seemed to represent the fabled missing link. The high-domed skull with its large brain cavity was similar to modern humans, yet the jawbone was apelike. Christened *Eoanthropus dawsoni*, Piltdown Man was attractive to contemporary palaeoanthropologists because it conformed to their expectations, suggesting that large brains evolved early in the human lineage, and that mankind itself had originated in Eurasia, which was more palatable than the African origin suggested by Darwin, for instance.

Further finds followed, confirming the importance of the site, including, in 1914, an extraordinary worked elephant-bone implement, shaped like an archaic cricket bat! Dawson died two years later, but the historical significance of his discovery seemed incontrovertible.

Yet the provenance of Piltdown Man was challenged from the start, with suggestions that the skull was too similar to modern man, and the jawbone too similar to a modern ape. Discoveries of genuine human ancestors in Africa and the Far East contradicted the evidence of Piltdown, with small-brained skulls, and detailed study of the skull by German scientist Franz Weidenreich led him to conclude it must be a fake. Although he could not prove it, he was sure it had been cobbled together from a human skull and an ape's jaw, saying, 'The sooner the chimaera "Eoanthropus" is erased from the list of human fossils, the better for science.'

It was not until 1953 that Piltdown Man was proved to be a fake, constructed by matching an artificially stained and modified human skull with the jawbone of an orangutan. It is now known that Dawson faked many other of his prehistoric and antiquarian 'finds' and, according to Dr Miles Russell, 'Piltdown was not a "one-off" hoax, more the culmination of a life's work.'

FRAUD IN SCIENCE

Science has proved to be a tool of enormous power for both explaining the world and transforming it. The source of this great power is the scientific method, a philosophy and set of practices that governs how science should be done: observations lead to hypotheses, which are tested and discarded if the evidence does not support them; if the hypothesis stands up it becomes a theory, but this in turn is tested through replication of the research and review by other scientists. One of the key strengths of the scientific method is that it is supposed to be proof against fraud or even error, and is frequently lauded as the sole road to Truth.

Scientific safeguards

Building on the ideal of the scientific method, modern science has built up a system of safeguards. The peer review system directs funding towards science and scientists deemed legitimate. The results of research are then presented in the form of papers submitted to journals, which are reviewed by 'referees', who judge the quality of the paper and decide whether it merits publication. The journals ensure that researchers present their methods and results in such a way that they can, theoretically, be checked and even, as the ultimate safeguard, replicated.

Thanks to this system, other scientists are supposed to be able to have faith in published findings, and they can then use these as the building blocks of their own research. This is how science progresses; scientists do not need to begin from scratch every time, proving things from first principles. Instead they can, in words that Isaac Newton adapted from an earlier author, see 'further by standing on the shoulders of giants'. But if scientists cannot trust the safeguards, the whole edifice of science comes under threat. Not only will the efforts of scientists going forward be hampered (as happened in British palaeoanthropology in the wake of the Piltdown Man forgery, when scientists dismissed the initial finds of *Australopithecus* skeletons because they seemed to contradict the evidence of Dawson's fake fossil), but there may also be genuine dangers on the applied side. Medical research is probably the most active field of scientific research, and false or erroneous findings could easily cost lives.

Types of fraud

In his 2004 book, *The Great Betrayal*, science journalist Horace Freeland Judson identifies three types of scientific fraud: fabrication, plagiarism and falsification. Fabrication is outright lying – creating observations or results with obvious intent to deceive. One example is the recent high-profile case of Hwang Woo Suk, the South

Clone or con? South Korean stem cell researcher Dr Hwang Woo-Suk – now disgraced – with Snuppy, the Afghan hound he claimed to have cloned from the dog on the left.

Millikan at work. Nobel Prize-winning physicist Robert Andrew Millikan, in his lab at Caltech where he discovered cosmic rays (and where he served as president for 24 years).

Korean cloning expert who claimed to have achieved a number of breakthroughs in his field, including the world's first cloned dog, Snuppy, and groundbreaking extraction of stem cells from cloned human embryos. Hwang became a national hero, international scientific celebrity and potential Nobel nominee, with papers published in the prestigious journal *Science*. But in 2005 his career fell apart after claims he had breached ethics rules and fabricated results; the police were brought in and his computer and files were impounded, and he was later charged with embezzlement for misuse of research funding. Another headline grabbing example was the Schön affair, in which garlanded young German physicist Jan Hendrik Schön, of America's prestigious Bell Labs, claimed to have created molecular computer transistors. Again, Schön was mooted as a potential Nobel recipient, until scrutiny of his research in 2002 showed that he had fabricated results in what one German academic called 'the biggest fraud in physics in the last 50 years'.

Plagiarism includes copying of both words and ideas (such as experimental design or results). Harvard professor and respected biologist Richard Lewontin argues that the widespread system of 'gift authorship', where senior scientists claim lead authorship of papers written by their students or junior colleagues, in many cases despite having had no involvement whatsoever, is a form of fraud: 'if laboratory directors as a matter of course claim authorship of work to which they have made no intellectual contribution or only a trivial one then they are, year in and year out, committing an intellectual fraud from which they reap immense rewards of ego, prestige, income, and social power.'

The third type of fraud, and perhaps the most widespread and insidious, is falsification. This is where results or observations from genuine experiments and research are tweaked, trimmed or otherwise adjusted, or simply ignored. There are many ways in which this can happen without intent to deceive and can even seem legitimate. For instance, in any experiment there are likely to be aberrant results, such as when a measuring device appears to be poorly calibrated, or a contaminant has entered the experimental system. It is only natural to wish to discard these 'nonsense' results. The best-known example is Robert Millikan's determination of the charge of the electron through experiments on tiny drops of oil, which earned him the 1923 Nobel Prize. Despite claiming to have published the results of 'all of the drops experimented on during 60 consecutive days … no single drop being omitted', it later transpired that he had excluded two-thirds of his original measurements.

A more subtle form of falsification is 'optional stopping' – this is where the experimenter has the option to stop counting or measuring, and does so when the results so far happen to match his initial expectations/desires. Then there is the 'bottom drawer' effect, in which studies with inconvenient or negative results tend to be 'locked away in the bottom drawer' and go unreported, whereas studies that prove a hypothesis are widely trumpeted. This is a particular problem in medical research, a field where commercial imperatives threaten to affect the integrity of the scientific method as much research is carried out or funded by industry.

How widespread is the problem? According to a 2005 survey of 3,427 scientists by the University of Minnesota and the HealthPartners Research Foundation, up to a third of the respondents had engaged in ethically questionable practices, from ignoring contradictory facts to falsifying data. Fraud is believed to be getting worse, partly because of the increasing number of scientific papers being published in a ballooning number of journals and partly because of the difficulties in policing globalized science.

FIEDEL

vs

DILLEHAY

FEUDING PARTIES
Stuart Fiedel (born 1952)
– archaeologist
vs
Tom Dillehay (born 1947) and
his Monte Verde team of
archaeologists

DATE
1999

CAUSE OF FEUD
Dispute over dating of Monte
Verde, a Palaeoindian site in
Chile, and of the first peopling
of the Americas

The last major landmasses to be settled by modern humans were the Americas, separated from the Old World by vast oceans, except at the extreme north-western tip where a land bridge between Alaska and Siberia disappeared and reappeared as sea levels fluctuated during the Ice Ages. The traditional view of the original peopling of the Americas is that the first significant human migration occurred via this land bridge during one of the periods when sea levels were low enough to expose the land bridge, while at the same time there was an ice-free corridor through the ice sheet covering Alaska and Canada.

Clovis people

The date of this colonization has been inferred from the presence of characteristic 'Clovis' fluted spear points at sites across the Americas, named for the site in New Mexico where they were first discovered in 1929. Bones and other evidence found with Clovis points have been dated to 11,500 years ago at the earliest, and this has long been accepted as the upper age limit for human presence in the New World. The conventional story is that a group of hunter-gatherers from East Asia migrated across the Siberia–Alaska land bridge and into the south, where lack of competition and abundant resources allowed a population explosion that led to the rapid spread of Clovis culture across much of the New World. This model is backed up by DNA evidence that suggests that all Native Americans are descended from East Asians, with the two populations diverging relatively recently.

Archaeological bombs

The Clovis model is undermined, however, by a number of sites that are claimed to be much older, including footprints found near Mexico City said to be 38,000 years old, and skulls found in the same region dated at 13,000 years. Silvia Gonzalez, the scientist behind both finds, recognizes that such interpretations are likely to be met with considerable resistance from mainstream American palaeoanthropology. Speaking to the BBC in 2005, she commented: 'It's going to be an archaeological bomb, and we're up for a fight.'

The most heavily studied of the alleged pre-Clovis sites, generally regarded as the strongest evidence, is Monte Verde in Chile, where charred wood, craft products, apparently worked

◀ *Clovis blade.*
Three views of a typical Clovis point, a spear blade characteristic of the dominant palaeoindian culture around 11,000 years ago.

'Fiedel's review is clearly biased...'

MICHAEL COLLINS, ARCHAEOLOGIST, 1999

stones and animal remains, all of which seem to indicate human presence and activities, have been carefully dated to around 12,500 years ago, a millennium before the Clovis culture. Recognizing that their work would be controversial, in 1997 the investigating team led by Tom Dillehay invited a commission of specialists to review the site; their conclusion was that Monte Verde did indeed predate Clovis.

Monte Verde

On the banks of a small river in a hilly region of southern central Chile lies what might be the most important Palaeolithic site in the Americas. Researchers believe that evidence of human habitation on the river's shores was preserved for millennia when the waters of the river rose, flooding the site and turning it into a peat-bog, a low oxygen environment in which organic materials can survive for long periods. It is thought that a longhouse of wood and hides originally stood here, separated into rooms by screens of similar construction, all held together by ropes made from local reeds. Clay-lined hearths would have formed the foci of communal life, and the remains of a variety of edible foods were found around these, including the oldest potato on record, seaweed presumably brought from the coast some 37 miles (60 kilometres) away, and other plants from outside the local area, which testify to the possibility of a trade network. Other finds included animal bones (including those of mastodons), stone tools, fossilized human faeces and a child's footprint.

'Troubling doubts'

Two years later, however, a review of the Monte Verde evidence by archaeologist Stuart Fiedel unleashed a storm of harsh words. He identified a number of errors, gaps and inconsistencies in the Monte Verde site reports, claiming that his findings raised 'troubling doubts' about the assertion that the site was indeed pre-Clovis. Members of the Monte Verde team were incensed; Michael Collins blasted Fiedel: '[his] review is clearly biased and negative in tone. He ignores material that does not support his critical thesis and takes the more negative or improbable of alternative views of each case that he discusses.' Tom Dillehay, the Monte Verde team leader, was even more scathing, complaining that, 'Fiedel obviously does not comprehend the nature of long-term, interdisciplinary research at a wet archaeological site like Monte Verde.'

The controversy still rages, but opinion is increasingly shifting to the view that there may have been at least one pre-Clovis colonization of the Americas, probably via a coastal route. Tribes expert at navigating shallow waters and living off coastal resources, such as shellfish and seals, could have followed currents and kelp forests from the eastern coast of Siberia across to north-western America, and indeed there does seem to be evidence of pre-Clovis settlement on the Channel Islands off the coast of California.

KENNEWICK MAN

The discovery of a 9,300-year-old skull at Kennewick in Washington State has stoked controversy on a number of fronts. Although he post-dates Clovis, Kennewick Man has a dolichocephalic skull, which means he has a long, narrow skull shape, in contrast to the shorter, broader skulls that modern Native Americans almost certainly inherited from their East Asian ancestors. He is thus seen as strong evidence that some humans did arrive in the Americas in a separate wave of colonization from the dominant Clovis culture – but how and from where?

Some argue that his skull shape is reminiscent of Australian aborigines, and that the first Americans were thus Australians who had somehow crossed the Pacific (the Polynesian settlement of the Pacific islands proves this is theoretically possible). Others – including some with a racist/white supremacist agenda – have used

▲ **The face of Kennewick Man.** *Reconstruction showing dolichocephalic head shape.*

Kennewick Man to argue that the first Americans were from Europe, and crossed the Atlantic thousands of years before the Vikings managed it. Kennewick Man has even been assigned to one of the lost tribes of Israel. Under the Native American Graves Protection and Repatriation Act, Native Americans in Washington have claimed Kennewick Man as one of their own but, because of the doubts about his ethnic origin, scientists have, so far, successfully retained access to the skeleton for study.

SMIT

vs

KELLER

FEUDING PARTIES
Jan Smit (born 1948) – Professor
of Earth Science at the Free
University of Amsterdam
vs
Gerta Keller (born 1945) –
Professor of Geosciences at
Princeton University

DATE
2001 onwards

CAUSE OF FEUD
What killed the dinosaurs?

Mass extinctions are a feature of the history of life on Earth and, although not the largest, the extinction event that saw the abrupt termination of the Age of the Dinosaurs is the one that captures the public imagination. For over 160 million years dinosaurs ruled the planet, until, in the geological blink of an eye, they vanished. This geological moment – the boundary between the Cretaceous and the Tertiary periods – is known as the K–T boundary (represented by an actual stratum in rocks of the period), and whatever killed the dinosaurs is known as the K–T extinction event. Every species of dinosaur except the birds seems to have disappeared at the K–T boundary, along with many other plants and animals, including plesiosaurs and marine phytoplankton.

Evidence of impact

The exact nature of the K–T extinction event was a mystery, but clues began to emerge. In 1979, analysis of the K–T stratum showed massively high levels of iridium, an element rare on Earth but present in asteroids, suggesting that a colossal meteorite had slammed into the Earth and vaporized, showering its iridium content all over the planet to settle as a thin layer. Further study showed the presence of a layer of spherules – tiny spheres of rock that form when molten rock is blasted into space, cools into tiny droplets and rains back down to Earth; a layer of soot, suggestive of massive fires presumably triggered by the impact and the rain of white-hot spherules; followed (in geological and therefore chronological terms) by high levels of fern spores. Massive fern growth is usually indicative of other plant life being wiped out.

A scenario was taking shape: an enormous meteorite, 6 miles (10 kilometres) across, had smashed into the planet around 65 million years ago. The space rock itself had vaporized and its impact with the crust had hurled millions of tonnes of ejecta into the atmosphere. Fire had ravaged the planet, possibly followed by acid rain and a post-impact winter in which the soot and dust thrown up by the impact screened the surface of the Earth from sunlight for months or even years. The dinosaurs and many other species failed to cope with the cataclysm and its after-effects, although some creatures, such as mammals, survived, perhaps burrowing beneath the surface to avoid the fire and cold.

An enormous impact might have been expected to leave an obvious crater, but none was known that fit this time period. Geologist Alan Hildebrand spent the 1980s

TIMELINE

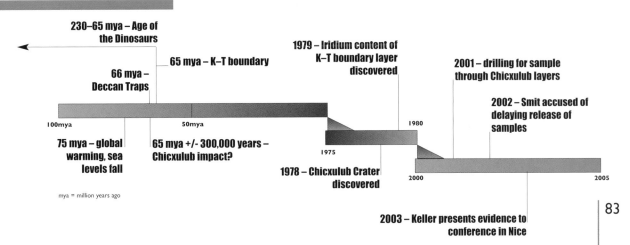

230–65 mya – Age of the Dinosaurs

66 mya – Deccan Traps

65 mya – K–T boundary

1979 – Iridium content of K–T boundary layer discovered

2001 – drilling for sample through Chicxulub layers

2002 – Smit accused of delaying release of samples

100mya

50mya

1980

75 mya – global warming, sea levels fall

65 mya +/- 300,000 years – Chicxulub impact?

1975

1978 – Chicxulub Crater discovered

2000

2005

mya = million years ago

2003 – Keller presents evidence to conference in Nice

83

searching for a candidate, tracking densities of spherules, which were thickest around the Caribbean. In 1990, he became aware of the discovery in 1978 by petroleum geologist Glen Penfield of an odd formation hidden in the geology of Mexico's Yucatán Peninsula. It was a 65-million-year-old crater centred on the town of Chicxulub, and it is now widely considered to be the scar left by the meteorite that killed the dinosaurs.

Stratigraphic squabble

A major proponent of the Chicxulub impact theory was Dutch geologist Jan Smit, who had uncovered the evidence of the spherules. Smit found stratigraphic evidence that appeared to illustrate the chain of events. Separating the spherule and iridium layers was a deposit of sandstone, which he interpreted as evidence of a titanic tsunami, a thousand feet high, which had scooped up vast amounts of sand and dumped them immediately on top of the initial layer of spherules laid down by the impact. The finer iridium dust took longer to settle and ended up on top of the sandstone, the whole process taking just a few days.

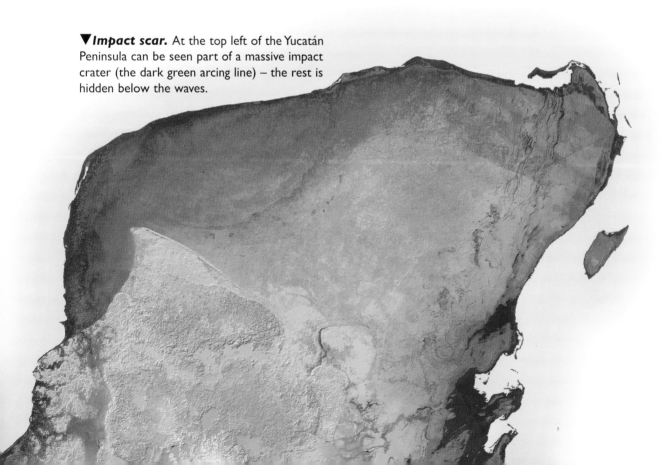

▼**Impact scar.** At the top left of the Yucatán Peninsula can be seen part of a massive impact crater (the dark green arcing line) – the rest is hidden below the waves.

'Gerta Keller's totally wrong.'

JAN SMIT, GEOLOGIST, 2004

Gerta Keller read the stratigraphic evidence very differently, however. Together with her colleague Wolfgang Stinnesbeck, she investigated similar sedimentary layers in the region and decided Smit's theory was wrong: 'It's a nice story but when you take a second look the story just doesn't stack up, there is something fishy about it, it just doesn't make sense.' She saw evidence that the sandstone layer took much longer than a few days to form: worm tracks and crab trails laid down over millennia; intrusive layers showing other rock formation periods; outcrops of fine-grained muddy material that would have taken long periods to settle; a base of limestone that could not have survived the acid rain supposedly associated with the impact; and, perhaps most tellingly, a secondary layer of spherules some 26 feet (8 metres) below the sandstone.

Looking at this lower layer gave Keller her eureka moment: 'I realized that if I'm right, everyone else is wrong; then science is wrong about one of the biggest catastrophes in history.' Her interpretation was that this lower, much older layer of spherules was the true footprint of the Chicxulub crater, while the higher, younger layers of spherules and iridium dust derived from another impact around 300,000 years later, the crater for which has yet to be found.

Core evidence

At this point things began to turn nasty. Smit has questioned Keller's interpretations and her ability to do science: 'when I think of Gerta Keller's so called evidence, it makes me sort of mad because none of it holds up. It's based on arguments which to my mind are barely scientific.' Keller accuses Smit of lashing out in defence of a wounded theory: '[He] has an awful lot at stake … [he's] desperate to rescue his impact tsunami hypothesis.'

In late 2001, in an effort to settle the issue, an oil-rig drill was used to drill down through the crater and obtain a core that would reveal the sequence of rocks and other

'Jan Smit says the things he does because he is desperate.'

GERTA KELLER, GEOLOGIST, 2004

evidence. Much of the core revealed shattered rocks, smashed into fragments by the force of the Chicxulub impact, but there was a solid layer between the pulverized rock and the iridium layer. In theory this layer could prove who was right: if it had formed almost instantaneously it would support Smit and his tsunami hypothesis; if it had taken a long period it would support Keller's double-impact hypothesis.

When the cores were sent straight to Smit for examination and redistribution to other interested parties, Keller and her supporters were appalled, and when Smit delayed the distribution for a year they were furious. 'We were dismayed,' wrote geochemist Erika Elswick of Indiana University in the journal *Nature* in 2003, 'There was no explanation given, no apology.' Keller had her own explanation. April 2003 would see a major conference of the European Union of Geosciences in Nice, France, and Keller claimed, 'He tried to postpone our results so that he could remain unchallenged at that meeting', an allegation Smit called 'ridiculous'. He claimed the delay was due to his busy schedule, and in fact Keller was able to examine a core sample before the conference, delivering a paper there outlining what she saw as conclusive evidence that she was right: a green clay mineral that her analysis identified as Glauconite, which requires extended periods of calm conditions to form; and fossil planktons of a species supposedly wiped out by the K–T event. If she was right the Chicxulub impact and the K–T event must be different.

Smit insists that her analysis was badly wrong. The fossil planktons were simply naturally occuring crystals over-interpreted through wishful thinking, and the Glauconite was actually Smectite, a mineral that can form very rapidly: 'So they made another mistake, they misidentified the clay mineral with very severe implications.' The drilling had solved nothing and the feud continues. 'We looked at all [the] evidence and I refute everything. There's not a single argument [of hers] which holds ground,' says Smit. 'He built his career on this impact-tsunami hypothesis and right now he's feeling this hypothesis crumble in front of his eyes,' shoots back Keller.

▲ **Artist's impression** of a meteor impact. The impact at Chicxulub in Mexico is the most popular candidate for the K–T event that wiped out the dinosaurs.

There is gathering evidence that they both might be wrong, with a more nuanced view of the K–T extinction linking the decline of the dinosaurs to global warming over several million years, accelerated by a huge burst of volcanism around the K–T boundary known as the Deccan Traps, when huge lava flows covered much of Western India, releasing enormous quantities of gas into the atmosphere. Meanwhile, some feel the Smit–Keller feud is not simply a distraction from what might be the real reasons for the decline of the dinosaurs, but actually damages science. According to Norman MacLeod of the Natural History Museum, London, 'This affair has become an object lesson on how partisan and unethical the whole dinosaur controversy has become. Young scientists are now refusing to get involved in this field because no matter what they say it will offend someone and damage their careers … No matter what you say, someone will hate you for it.'

JACOB

vs

BROWN, MORWOOD, ROBERTS, et al.

FEUDING PARTIES
Teuka Jacob (1929–2007)
– palaeontologist
vs
Peter Brown (born 1954),
Mike Morwood (born 1950),
Bert Roberts (born 1959), et al.,
of the Australian–Indonesian team
investigating Liang Bua cave on
the Indonesian island of Flores

DATE
2005 onwards

CAUSE OF FEUD
Dispute over existence of *Homo floresiensis*

The biggest development in palaeoanthropology in the last decade has also occasioned the biggest spat. In 2004, in a paper in the prestigious journal *Nature*, a team of Australian and Indonesian scientists announced an astonishing find. Bones and a skull recovered from a limestone cave on the Indonesian island of Flores appeared to prove that a previously unknown species of human, named *Homo floresiensis* but more popularly nicknamed the Hobbit because of its small stature, had lived there until as recently as 13,000 years ago.

Apart from ourselves, these Hobbits would have been the only species of the *Homo* genus to have survived into relatively recent times, and would have co-existed with *Homo sapiens* for millennia. The find lent startling substance to speculation about Indonesian apemen, such as the legendary *orang pendek*, and to theories about the origins of 'little folk' ubiquitous in folklore around the world. It also threatened to seriously undermine the multiregional hypothesis of human origin and bolster the arguments of the Out of Africa crowd (*see* pages 62–65), with the result that the Hobbit was pitched headlong into the centre of a vicious scientific brouhaha.

Kidnapped!

The Hobbit team based their announcement of a new species on the bones of eight individuals, including a single skull with a cranium just one-third the size of a modern human. An initial hypothesis was that *Homo floresiensis* was descended from a population of *Homo erectus* who had settled the island at least 80,000 years ago, and who, in their isolated habitat, had evolved into a much smaller form, but several features of the skeletons mitigate against this and seem to show the hobbit lineage diverged from that of other *Homo* species over a million years ago. Somehow they reached Flores, where they co-existed with, but were possibly eventually wiped out by, the distinctly different species *Homo sapiens*. This is the opposite of the scenario envisaged by multiregionalists, who argue that *Homo* populations in different locations interbred, allowing them to remain a single species while collectively evolving into modern humans. Perhaps it was not surprising that Professor Teuka Jacob, the doyenne of Indonesian palaeoanthropologists and a staunch multiregionalist, took umbrage at such a radical blow to his favoured theory emerging from his own backyard.

In 2004 the 'Hobbit' bones arrived at Jacob's laboratory in Jakarta, where he set about disproving the *Homo floresiensis* hypothesis.

▲ *Little and large.* Skulls of *Homo floresiensis* (L) and *Homo sapiens* (R) side by side for size comparison.

◀ *Professor Mike Morwood,* one of the leaders of the team that discovered the Hobbit remains on Flores, displaying the tiny skull at the centre of the storm.

Bad blood surfaced immediately, as fears were voiced that Jacob had 'kidnapped' the bones and would restrict access to them. Richard 'Bert' Roberts, one of the co-authors of the *Nature* paper, warned darkly, 'Jacob has a habit of hanging on to fossils for a long time. He cannot be allowed to keep these, to stifle the study that he so advocates.' It was even suggested that Jacob had damaged the bones. He vigorously disputed the charges; the bones were sent to him, he claimed, by the Hobbit team themselves – 'They even gave me the money for the transport' – while in March 2005 he returned them, safe and sound.

Scientific terrorists

By now Jacob had formed strong opinions about the Hobbit team and their identification of the bones, which he argued were simply those of pygmy humans, found widely in the same region of Indonesia. The small skull he explained as a case of *microcephaly*, a condition in which the brain does not develop properly; it was simply the remains of what Jacob's friend and supporter, multiregionalist Alan Thorne, describes as 'a diseased pygmy'.

Jacob had harsh words for the Australians, claiming, 'They did their study without comparative material,' and calling them 'scientific terrorists' for forcing their ideas on people. 'I don't think the Australians have the expertise. They were very narrow. They have a tunnel vision and were not equipped in this area … I would say [to them] "do some more work. Think twice. Look at everything from different angles. Don't start with the conclusion.'" He even expressed doubts about the scientists that *Nature* had chosen as the referees of the original article: 'The reviewers seemed unevenly selected, very one-sided.'

Cushioned arses and sour grapes

The 'Australians' shot back. Roberts categorically rejected the charge about the referees: '[*Nature*] had six referees on each paper, the most I have ever known. They made damn sure they had a cushion behind their arse.' As for his team: '[We] had everyone involved – geomorphologists, geochronologists, archaeologists, paleoanthropolgists ... We left no bone unturned. Good grief, it was a soccer team of authors!' The real motivation of Jacob and other critics, Roberts said, was to protect their favoured model of evolution: 'All ... are supporters of multiregionalism ... This discovery would destroy their theory. It suits their purposes very nicely [to oppose *Homo floresiensis*].'

Peter Brown, who led the Hobbit team, was more blunt. He dismissed as 'complete rubbish' and 'sour grapes' Jacob's contention that '[the specimens] could just be Pygmies ... of course, there are small-bodied people on Flores,' Brown countered, 'but they don't have brains one-third the size of ours, or unusually shaped pelvises or very long arms like *Homo floresiensis*.'

Since this initial spat, further controversy has raged over the Hobbit skull, with researchers hurling insults and disputing conclusions. As Henry Gee, an editor at *Nature* who championed the initial paper, has pointed out: 'Science is a disputatious business, and human evolution is notorious for being even more disputatious. Historically, whenever anyone discovers a new hominid, a lot of people come along and say it's an ape or a diseased human.' He admits '[The Hobbit team] are going to have to discover some more bones that prove this,' but insists that 'we have history on our side.'

In fact, history shows we are doomed to repeat its mistakes. In the 14th century, Marco Polo described how the natives of Sumatra would try to sell the mummified bodies of pygmies to visitors as souvenirs. Today, we know that pygmies do indeed inhabit these islands, but Polo, sounding much like a latterday Jacob, was having none of it. ''Tis all a lie and cheat,' he declared. 'Those ... little men ... are manufactured on the island.'

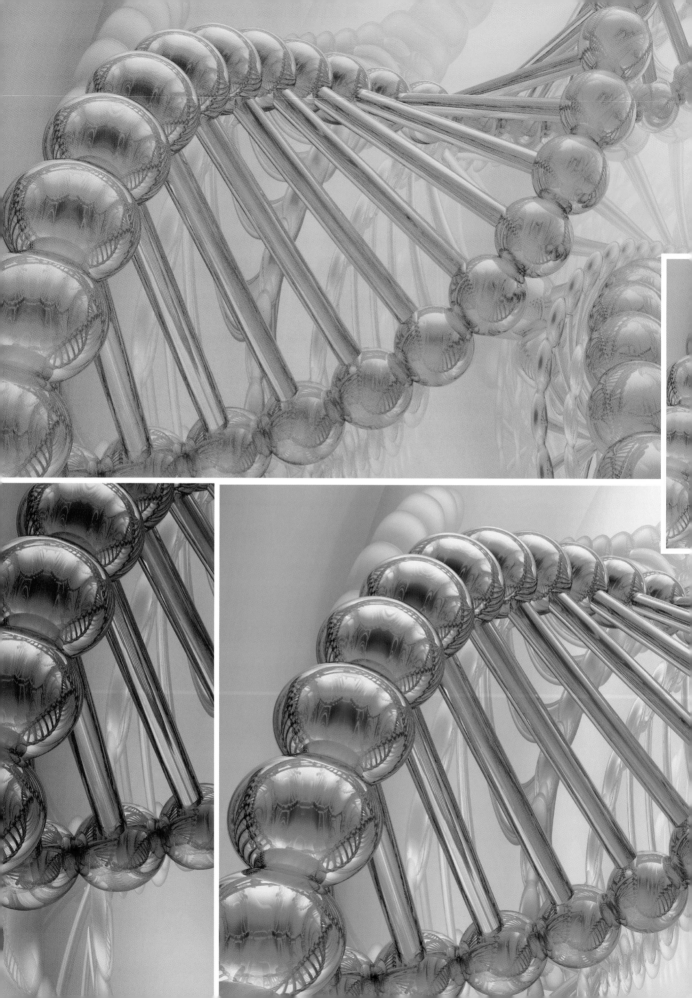

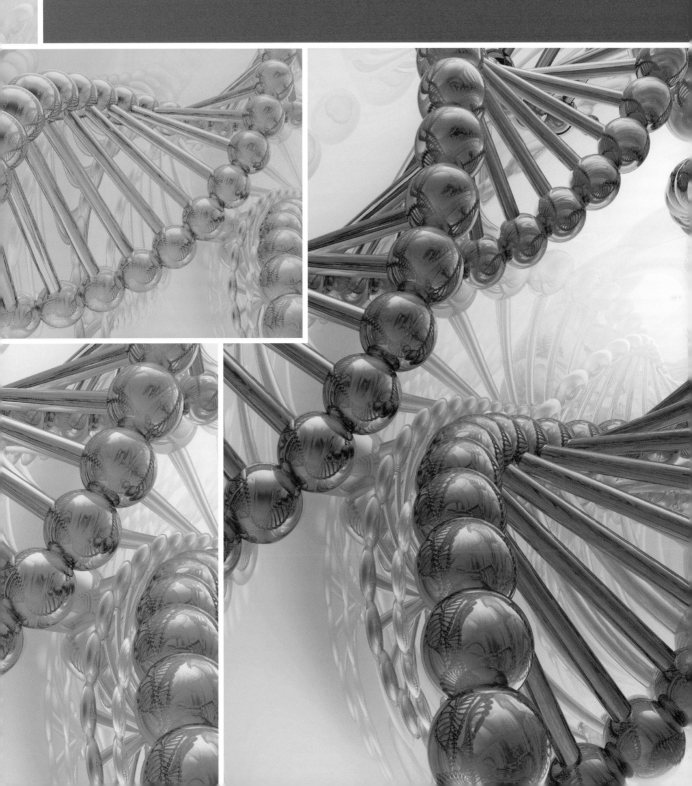

PART THREE
BIOLOGY
and MEDICINE

HARVEY

vs

PRIMROSE
and RIOLAN

FEUDING PARTIES
William Harvey (1578–1657)
– physician
vs
James Primrose (1600–1659);
Jean Riolan the Younger
(1580–1657) – physicians

DATE
1628–50s

CAUSE OF FEUD
Circulation of the blood

In 1628, the English physician William Harvey published a revolutionary work, *Exercitatio de motu cordis et sanguinis in animalibus*, more generally known as *De Motu Cordis* – 'On the motion of the heart'. In it, he laid out the careful experimental work that had convinced him to overturn 1,400 years of medical orthodoxy and contradict the teachings of the Greek physician Galen (*c*.130–201CE), taken as law by most physicians ever since.

Harvey versus Galen

In the Galenic tradition, the liver converted food into blood, which was heated in the heart where it was imbued with vital spirits, and then seeped out through the blood vessels to nourish the body. Harvey found out for himself through dissections, vivisection and other experiments that not only was the heart a vigorous pump, but also that the volume of blood that passed through it meant that there could not be enough in the body to stop it from running empty, unless it was being circulated.

Where Galen had taught that the cardiac septum – the muscular wall separating the ventricles of the heart – was pierced by pores, allowing blood to seep from one side to the other, Harvey found that this was not the case: 'But, damme, there are no pores, and it is not possible to show such.' Instead, he posited – but could not prove – that there must be a network of tiny blood vessels connecting the arteries that carried blood away from the heart to the veins that carried blood back towards it, completing the circuit.

'All men as enemies'

Harvey knew that his new theory might provoke controversy on many fronts, writing in *De Motu Cordis*: 'What remains to be said on the quantity and source of the transferred blood, is, even if carefully reflected upon, so strange and undreamed of, that not only do I fear danger to myself from the malice of a few, but I dread lest I have all men as enemies, so much does habit or doctrine once absorbed, driving deeply its roots, become second nature, and so much does reverence for antiquity influence all men.'

Among the first to voice criticism of Harvey was the young physician James Primrose, who desired to make a name for himself by attacking the popular new theory. In a 1630 tract called 'Exercitations and animadversions' – according to historian of medicine Roger

▶ **William Harvey** *trained in medicine at Cambridge and Padua, Italy, where he was influenced by important anatomists such as Vesalius and Fabricius, who stressed the importance of first-hand investigations of human*

▲ *A 19th-century engraving* showing Harvey demonstrating his theory of circulation of the blood to King Charles I with the help of a deer.

French, 'conceived as a way of publicizing himself' – Primrose argued that Harvey's doctrine was little more than a fashionable novelty, which threatened to destroy traditional medicine. Primrose objected to Harvey's calculations of cardiac output, despite having performed no experiments of his own, and complained that Harvey's admirers treated his words 'as if they came from an oracle'.

Most of Harvey's critics came from the old guard, whereas younger physicians generally supported him. Why was Primrose different? Historian Geoffrey Keynes suggests that Primrose was 'congenitally unable to accept new ideas' but his motivations may have been more complex. French speculates that Primrose's antipathy to Harvey's book may have been triggered by a spat with John Argent, president of the College of Physicians, where 'after a sharp exchange on the likelihood that Harvey was right, Primrose was put firmly in his place by Argent and told to go away and read Harvey's case.' Another theory is that Harvey had made an enemy in Primrose because he had been one of his examiners in the College of Physicians, the same college that had acted to prevent Primrose from lecturing and barred him from admittance on the basis that he was a foreigner (though his father was a Scot, he had been brought up in France).

The old guard

When Harvey visited Europe to lecture on his new theory, he ran up against Caspar Hofmann, a professor of medicine at the University of Altdorf in Germany. In 1636, Harvey demonstrated circulation for him through vivisection, but Hofmann remained unmoved. Hofmann believed that circulation would 're-cook' blood into bile, and that in trying to measure cardiac output, Harvey was daring to 'to quantify the unquantifiable'. He also pointed out one of the most serious flaws in Harvey's theory: it depended on the existence of capillaries, but these were not discovered until 1660.

Harvey's greatest opponent, however, was the French physician Jean Riolan the Younger, a staunch advocate of the ancient traditions of medicine. French explains: 'Riolan saw himself in a tradition of medicine stretching from Cos [ancient home of Hippocrates] to Paris, in which too his father had been a part.' According to historian of medicine Andrew Wear, '[Riolan] tried to save Galenic medicine by limiting the effect of Harvey's circulation.' Riolan's arguments against Harvey were similar to those of Primrose and Hofmann, and he was particularly worried because the new doctrine undermined the rationale for bleeding, a major therapeutic tool at the time, although it is notable that doctors continued to bleed hapless patients for centuries afterwards.

Eventually, the mild-mannered Harvey was prompted to respond to Riolan, complaining about his refusal to believe experimental evidence: 'Who will persuade a man that has not tasted them, that sweet or new wine is better than water? With what arguments shall one persuade a blind man that the Sun is clear, and outshines all the Stars in the firmament? So concerning the Circulation … .'

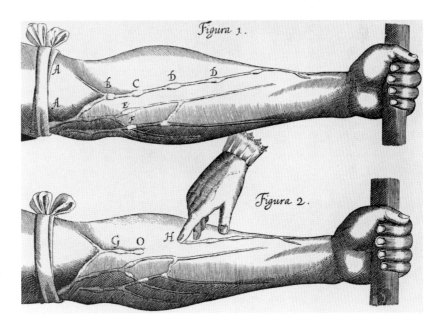

▶ **Pressure points.** One of Harvey's experimental proofs was a demonstration of the action of valves in the veins of the arm.

SCIENCE AND EXPERIMENTAL EVIDENCE

Science was born out of a kind of feud; a clash between competing philosophies of thought and the proper route to truth (a field known as epistemology). This epistemological collision was between a philosophy that had ruled Western thought throughout the Middle Ages and had its roots in Ancient Greece, and a new philosophy that would later become known as science. One of the key weapons in this battle was itself a matter of contention: the experiment – 'a controlled manipulation of events, designed to produce observations that confirm or disconfirm one or more rival theories or hypotheses', according to *The Oxford Dictionary of Philosophy*.

Crown all other learning

The era that saw the birth of science is known as the Early Modern period. Learning at this time was based on the wisdom of the ancients in tandem with the authority of scripture and the Church, a philosophy known as scholasticism. Above all, scholasticism looked to the ancient Greek philosopher Aristotle as not only the source

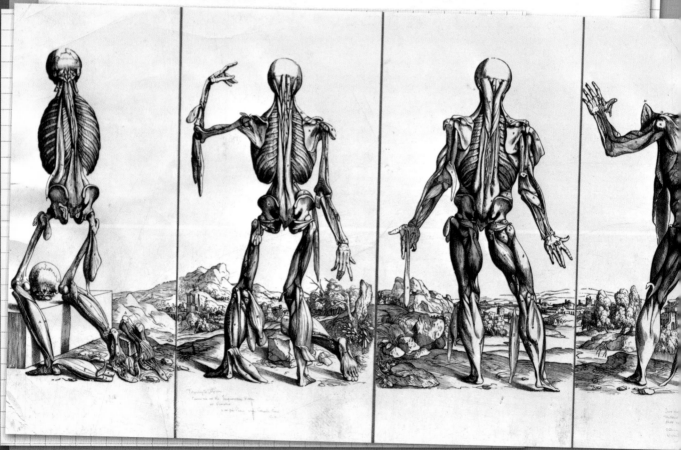

of knowledge but also the way to arrive at knowledge. A 17th-century study guide, 'Directions for a Student in the University', claimed 'The reading of Aristotle will not only conduce much to your study … but allso help you in Greeke, & indeed crown all your other learning.'

The scholastics believed that the only route to truth was through pure reason, in the form of deductive logic: for example, starting with certain assumptions about the natural world, logic could build on these to reach certain conclusions. For instance, Aristotle began with the assumption 'Everything that is in motion must be moved by something', and from this deduced that there must be a chain of transmission of motion that stretched back to the prime mover (for the medieval Church this obviously meant God). In all areas of knowledge the authority of the ancients was to be respected; therefore, astronomy still followed the precepts of the ancient Alexandrian Ptolemy (who taught that the Earth was the centre of the universe), and medicine followed the precepts of Hippocrates and Galen. Experimentation was looked down upon and actively discouraged; mere 'mechanics' and 'artificers', with their roots in trades such as mining and masonry, threatened to sully pure reason.

There were many problems with this scholastic epistemology. Deductive logic could only be relied upon to produce 'true' conclusions if the premises upon which it was operated were also true. But where did these premises come from in the first place? The senses could be misled, or could easily miss things. Relying on earlier authorities, or what was generally 'known' or 'accepted', was equally problematic.

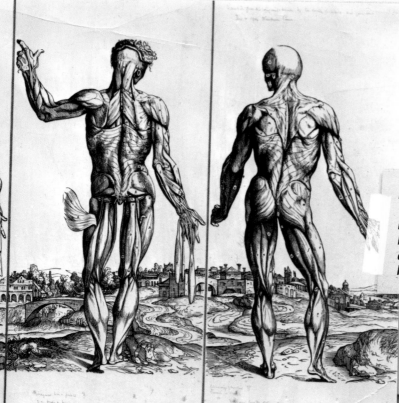

Muscle man. *The Italian anatomist Vesalius revolutionized his field by investigating human anatomy for himself, rather than relying on the word of the ancients. Striking illustrations helped to popularize his discoveries.*

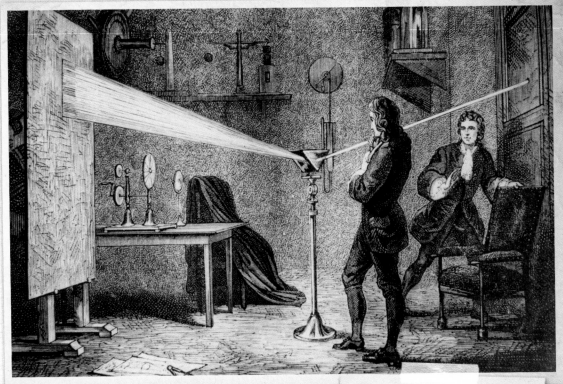

'We are certainly not to relinquish the evidence of experiments for the sake of dreams and vain fictions of our own devising'

Sir Isaac Newton, natural philosopher, 1687

A new philosophy

Gathering dissatisfaction with the limitations of the scholastic approach led to a new breed of natural philosophers intent on challenging authority and who stressed the need for philosophers to find out things for themselves. One of the most influential was the English statesman and philosopher Sir Francis Bacon (1561–1626), who argued that the goal of natural philosophy should be the interpretation of nature, not what he called 'anticipations of nature'; that scientists should begin with observation, avoiding preconceptions and unfounded assumptions, proceed by experiment and use inductive reasoning. One of his aims was 'to recall natural philosophy from the vanity of speculations to the importance of experiments'. According to legend, Bacon was killed in the course of an experiment on 'the conservation and induration of bodies', catching a fatal chill while attempting to stuff a chicken with snow in order to preserve it.

By the time Bacon was writing, the anatomist Andreas Vesalius (1514–64) had already challenged the Galenists. His dissections of human corpses had demonstrated that Galen was frequently wrong about anatomy, and thus changed the way in which medicine was taught (up to this point Galen had been presented as dogma, with no one checking his writing against reality). Harvey's work on circulation continued this trend, using experiments to challenge the received wisdom (*see* pages 94–97). Meanwhile, in physics, the pioneering experiments of Galileo challenged Aristotelian dogma on motion and gravity. Galileo's experiments with balls and inclined planes showed, for instance, that objects fell at the same rate irrespective of their nature and that an object's speed depended on the duration of its fall. Galileo's challenge to scholasticism was to culminate in his battle with the Church (*see* pages 166–171).

'Take nobody's word for it'

Bacon's experimental philosophy was to bear its greatest fruit in England, where in 1660 a group of 'mechanical philosophers', including Robert Boyle (1627–91), founded a society initially described as a 'Colledge for the Promoting of Physico-Mathematicall Experimentall Learning'. Two years later it was granted royal approval and became the Royal Society, crucible of the scientific revolution and an institution with the motto *nullius in verba* – 'take nobody's word for it' – an exhortation for members to find things out for themselves through experiment. In pursuing this approach, the new philosophers improved or invented new tools for experimentation, including the microscope, the thermometer and the air-pump.

The greatest of these new philosophers was Isaac Newton, arguably the first man to achieve the synthesis of observation, experimentation and mathematical proof that best characterizes the scientific method. From an early age, Newton set himself to overcome the limits of scholasticism, writing in a student notebook the epigraph, *Amicus Plato amicus Aristoteles magis amica veritas* – 'Plato is my friend, Aristotle is my friend, but truth is my greater friend'. He was always very concerned in his writing to stress that relying purely on the senses and/or 'mere' speculation was not the way to get results: 'The nature of things is more securely and naturally deduced from their operations one upon another than upon the senses.'

Newton also hated disputation (*see* page 179), and was at pains to point out that the theories he arrived at were not simply his opinion. For instance, in a landmark paper on light and colour, he insisted: 'For what I shall tell concerning them is not an Hypothesis but most rigid consequence, not conjectured by barely inferring tis thus because not otherwise ... but evinced by the mediation of experiments concluding directly & without suspicion of doubt.' He reserved particular scorn for what he called 'hypotheses' (in this context meaning speculations unsupported by experimental evidence), expressed in his famous lines: 'I frame no hypotheses; for whatever is not deduced from the phenomena is to be called an hypothesis; and hypotheses ... have no place in experimental philosophy.'

PASTEUR

vs

POUCHET

FEUDING PARTIES
Louis Pasteur (1822–95)
– chemist, microbiologist
vs
Félix Archimède Pouchet
(1800–1872) – naturalist

DATE
1860–64

CAUSE OF FEUD
Spontaneous generation

Spontaneous generation was an idea with a long history (*see* box, page 107) when Félix Archimède Pouchet, Director of the Natural History Museum in Rouen, France, came to it in the 1850s. Pouchet was determined not only to show how spontaneous generation happened but also to free it from its baggage of atheistic materialism. In a series of papers beginning in 1855, Pouchet laid out a theory of what he called *heterogenesis*. In 1859, he published a book of the same name. Pouchet agreed with one of the central claims of the anti-spontaneous generation camp: that a complex life form must develop from a spore, egg or some other simple embryonic form. In other words, all life might be generated from eggs (as William Harvey had proposed – *see* box, page 107) but he opposed them in asserting that eggs themselves might be generated spontaneously. For Pouchet, spontaneous generation was a property of organic matter, which, with the correct combination of air, water and temperature, could produce embryonic or egg-forms of life – 'a completely spontaneous act by which the plastic force brings together in a special organ the primitive elements of the organism.' In Pouchet's vision, this process depended on divine power, both at the beginning – the creation of life – and as an ongoing process. He insisted that: 'The law of heterogenesis, far from weakening the attributes of the Creator, can only augment Divine Majesty.' Pouchet claimed to have proved the reality of heterogenesis with a series of experiments in which meticulously sterilized water in meticulously sterilized flasks became filled with microorganisms despite all these precautions against contamination.

▼ **Pasteur in his lab.** *Not just a brilliant scientist, the legendary microbiologist was also adept at navigating the worlds of academia, industry and politics.*

13ᵉ ANNÉE — N° 612   PARIS ET DÉPARTEMENTS 15 CENTIMES   13 MARS 1886

LE DON QUICHOTTE

Rédacteur en Chef: Ch. GILBERT-MARTIN

BORDEAUX
Bureau : RUE CABIROL, 7
ABONNEMENTS
UN AN .... 10 fr.
SIX MOIS .... 5 .
L'ÉTRANGER LE PORT EN SUS
PARIS
DÉPÔT GÉNÉRAL ET VENTE
17, Rue Saint-Roy
Distribution dans les kiosques
Chez les libraires et les marchands
de journaux.

ANNONCES
LES ANNONCES SONT REÇUES
A L'AGENCE HAVAS
POUR LA PUBLICITÉ DE BORDEAUX
Péristyle du Grand-Théâtre
côté sud.
La ligne
Annonces sur 6 colonnes 25 c.
Réclames sur 5 colonnes 40 c.

L'ANGE DE L'INOCULATION (M. PASTEUR), par GILBERT-MARTIN.

'The doctrine [of spontaneous generation is] gratuitous and uncalled for'

JOSEPH LISTER, SURGEON, 1869

Eyes on the prize

The French Académie des Sciences had instituted the Alhumbert Prize of 2,500 francs, to be 'given to him who by well-conducted experiments throws new light on the question of so-called spontaneous generation'. This attracted the attentions of Louis Pasteur, a brilliant scientist with a keen appetite for money. Pasteur was not a fan of spontaneous generation, accusing its proponents of pseudoscientific sophistry. 'One must not assume that an understanding of science is present in those who borrow its language,' he remarked witheringly. Writing to Pouchet, Pasteur declared, 'I think Monsieur, that you are mistaken, not so much to believe in spontaneous generation, for in such a question it is difficult to avoid preconceived ideas, but to assert that spontaneous generation exists.'

Pasteur suspected that Pouchet's results were the result of contamination from the mercury that Pouchet had used to cool them, which collected dust (and therefore microscopic spores or eggs). Pasteur devised a series of flasks that exposed sterilized broth to air (which Pouchet insisted was necessary for heterogenesis) but prevented any dust contamination; the liquid in these flasks remained clear and free from germs for months. Clearly, Pasteur reasoned, the germs must be derived either from inanimate dust particles (for example, soot) or from spores or eggs and, favouring the latter reasoning, said 'I prefer to think that life comes from life rather than from dust.'

Pasteur wrote up his findings in an essay, 'Mémoire sur les corpuscules organisés qui existent dans l'atmosphère', published in 1862, which duly won him the Alhumbert Prize. According to Everett Mendelsohn, Professor of History of Science at Harvard, '[Pasteur] was contemptuous in tone of the work of his elder compatriot' – Pasteur was 23 years younger than Pouchet.

◀ **Pasteur to the rescue.** *His most successful work was the development of vaccines for diseases including rabies, and he was lionized in the French media – here he is portrayed as 'the Angel of Inoculation'.*

Subserving the prepossessions

According to the traditional account, Pasteur had won fair and square. His experiments were, according to Mendelsohn, 'by any standard … innovative and brilliant', while according to historian of science Nils Roll-Hansen, 'In contrast to his opponents, Pasteur was not led astray by his prejudices, but built methodically on experimental work done largely by himself.' Pouchet did not see it this way, however, and accused the commission that had awarded the prize of bias. The commission was disbanded and another one established in 1864, 'but once again the contestants could not even come to agreement on procedures or the open-mindedness and fairness of the committee members', explains Mendelsohn.

This was a question of science, yet neither the participants nor the committee set up to referee them could agree on basic standards of fairness. The differences between the two men, writes Mendelsohn, 'lay outside the science as well as within it'. The spontaneous generation debate was too wound up with politics to remain a purely scientific matter. The Académie was subordinate to the desires of the Emperor, Napoleon III, who was 'widely recognized as a vigorous opponent of republicanism, atheism, and materialism', and, as the British palaeontologist Richard Owen observed in 1868, 'Pasteur … had the advantage of subserving the prepossessions of the "party of order" and the needs of theology.'

Pasteur was expert at cultivating the patronage of the imperial family, and, according to Roll-Hansen, many modern historians of science are 'persuaded that external factors influenced Pasteur's research and scientific judgement more powerfully than they did the defeated Pouchet.' Shocking evidence of this emerged in a controversial 1995 book by Gerald Geison, *The Private Science of Louis Pasteur*. Geison had studied Pasteur's laboratory notebooks and discovered that he had suppressed the vast majority of his experimental results because they did not serve his cause; only 10 per cent of his trials gave the desired result (*see* page 77). Of course, this was not known at the time, and Pasteur was celebrated as a hero by the French for putting the sword to the atheistic canard of spontaneous generation (a view that ignored Pouchet's piety). Further afield, however, the spontaneous generation debate continued. Mendelsohn points out: 'the Pasteur–Pouchet debate … had to be repeated only a few years later in England when T.H. Huxley and John Tyndall took on Henry Bastian in a British version of the more famous French confrontation.'

CHICKEN OR EGG?

The question 'Which came first – the chicken or the egg?' perplexed the earliest philosophers. If organisms can only be generated as the result of reproduction by parents, where did the first in each line of organisms come from? Some philosophers assumed that life had always been in existence, others that some special act of creation must have been involved – i.e. by God. The ancient Greek philosopher Anaximander (611–547 BCE) cut the Gordian knot by proposing that the chicken (or the egg) could have originated spontaneously by the action of light on water. Aristotle elucidated the theory to the highest degree – 'Animals and plants come into being in earth and in liquid because there is water in earth, and air in water, and in all air is vital heat so that in a sense all things are full of soul. Therefore living things form quickly whenever this air and vital heat are enclosed in anything.'

Aristotle's take on spontaneous generation became the standard model in the West, and it was accepted that, just as he had said, everything – from frogs and flies to mites and moths – could arise out of some combination of matter. The advent of scientific investigation, however, and particularly the increasing power of microscopes, progressively depleted the range of creatures thought to arise from nowhere. The physician William Harvey (see pages 94–97), in his De Generatione of 1651, asserted that ex ova omnia ('all [life comes from] eggs') and identified invisible airborne 'seeds' as the source of many cases of apparent spontaneous generation. In 1668, the Italian physician Francisco Redi (c.1629–97) showed that maggots come from the eggs laid by adult flies. By the time the Pasteur–Pouchet controversy was settled, the scope of spontaneous generation was limited to the lowliest forms of primordial slime; and by 1869, the English surgeon Joseph Lister (1827–1912) was able to observe with disdain that the doctrine of spontaneous generation had been 'chased successively to lower and lower stations in the world of organized beings'. By the mid-19th century, spontaneous generation had become associated with materialism, atheism and evolution, to the point where adhering to its doctrine had become a political act, and its proponents were regarded with suspicion by the reactionary forces of religion and the state, especially in Second Empire France.

FAULDS

vs

GALTON
and HERSCHEL

FEUDING PARTIES
Dr Henry Faulds (1843–1930)
– medical missionary;
Colin Beavan (born 1964)
– author of *Fingerprints: The Origins of Crime Detection and the Murder Case that Launched Forensic Science*
vs
Sir Francis Galton (1822–1911)
– bio-statistician, polymath;
Sir William Herschel (1833–1917)
– judge;

Gavan Tredoux (born 1967)
– biographer of Galton and maintainer of galton.org

DATE
1894 onwards

CAUSE OF FEUD
Credit for pioneering forensic use of fingerprints in criminal detection

The realization that fingerprints are a unique and invariable marker of identity and the development of a system for recording, classifying and comparing them, ranks as one of the great landmarks in the history of crime detection. The row over who deserves the credit for this advance dates back to its inception and continues to this day, thanks mainly to Henry Faulds and his partisans who have continued to advocate his cause long after his death; recent advocates include author Colin Beavan, whose popular and well-received 2001 book *Fingerprints* has rekindled interest in Faulds' claims.

'The scientific identification of criminals'

Scientific investigation of fingerprints dates back to the 17th century, while the use of finger-impressions for purposes of identification may date to 14th-century Persia or even ancient Babylon. The realization that fingerprinting could be a valuable forensic tool, and a workable system for applying this knowledge, came later. In the mid-19th century William Herschel was a young magistrate in India, struggling with the problem of widespread identity fraud in administering state-issued pensions. He hit upon the idea of getting the applicant to make hand and finger prints on official documents – to begin with, almost as a sort of sympathetic magic, 'to frighten [him] out of all thought of repudiating his signature' through instituting personal contact with the document.

Herschel eventually realized that the fingerprint patterns could be useful identifiers, and went on to use them in other aspects of the Indian administration as he climbed the grades, becoming a judge in the Hooghly district in 1877 and instituting an official system of fingerprinting. After more than 20 years of collecting prints he had quite a collection, and was able to observe for himself, by reprinting individuals over time, that the prints remained invariate and identifiable.

Meanwhile, in Japan a forceful Scottish doctor and missionary, Henry Faulds, was reaching his own conclusions about fingerprints. He became interested in fingerprints in

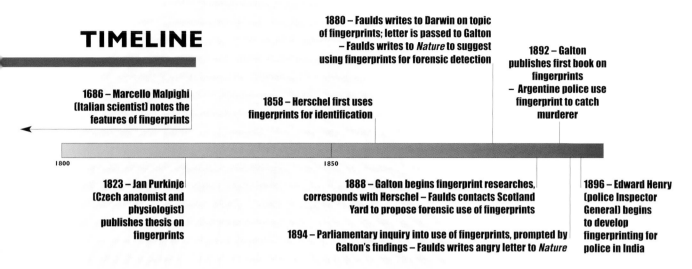

1880 – Faulds writes to Darwin on topic of fingerprints; letter is passed to Galton – Faulds writes to *Nature* to suggest using fingerprints for forensic detection

1892 – Galton publishes first book on fingerprints – Argentine police use fingerprint to catch murderer

1686 – Marcello Malpighi (Italian scientist) notes the features of fingerprints

1858 – Herschel first uses fingerprints for identification

1800

1850

1823 – Jan Purkinje (Czech anatomist and physiologist) publishes thesis on fingerprints

1888 – Galton begins fingerprint researches, corresponds with Herschel – Faulds contacts Scotland Yard to propose forensic use of fingerprints

1896 – Edward Henry (police Inspector General) begins to develop fingerprinting for police in India

1894 – Parliamentary inquiry into use of fingerprints, prompted by Galton's findings – Faulds writes angry letter to *Nature*

around 1879, after examining 'some specimens of prehistoric pottery' and noticing 'certain finger-marks which had been made on them while the clay was still soft'. After this, he claimed to have done a number of alarming experiments on his own fingers, including burning, scouring and soaking them in acid to see whether his prints could be erased or altered. In 1880, he wrote to the journal *Nature* 'On the Skin-Furrows of the Hand', briefly describing some of the features of fingerprints, the best methods to obtain a clear print, and, crucially, suggesting that 'When bloody finger-marks or impressions on clay, glass, &c., exist, they may lead to the scientific identification of criminals.' He even noted the 'for-ever unchangeable' nature of 'finger-furrows'.

Earlier that year, Faulds had written to Charles Darwin, hoping to attract the attention of the great man, but the ageing naturalist had passed on the letter to his cousin, Francis Galton, a man he knew to be interested in matters of biological measurement (*see* box, page 112). Beavan contends that Galton deliberately suppressed Faulds' letter and, later, maliciously denied knowledge of it; in fact, Galton appears to have taken little notice of it, passing it on to the Royal Anthropological Society and forgetting about it.

A matter for the police

In 1888, Faulds wrote to the Metropolitan Police in London to suggest the use of fingerprints for forensic purposes, but nothing came of the suggestion. Galton's primary advocate today, biographer Gavan Tredoux, suggests, 'The police may have considered Faulds a harmless crank, an impression that might have been reinforced by

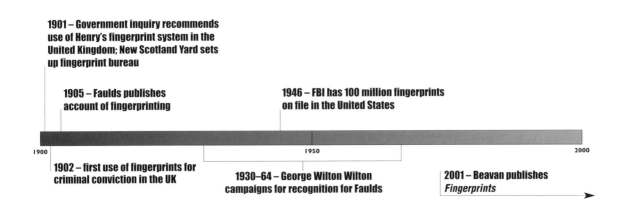

1901 – Government inquiry recommends use of Henry's fingerprint system in the United Kingdom; New Scotland Yard sets up fingerprint bureau

1905 – Faulds publishes account of fingerprinting

1946 – FBI has 100 million fingerprints on file in the United States

1900

1950

2000

1902 – first use of fingerprints for criminal conviction in the UK

1930–64 – George Wilton Wilton campaigns for recognition for Faulds

2001 – Beavan publishes *Fingerprints*

his aggressive personality.' That same year, however, a more consequential intellect was brought to bear on the topic of fingerprints. In 1888, Francis Galton was asked to give a lecture on Bertillonage (anthropometry), a system for recording physical characteristics of criminals for forensic purposes devised by French policeman Alphonse Bertillon.

Researching the lecture, Galton's interest was piqued. He realized that if fingerprints were to be accepted as a legitimate forensic tool he would have to prove that they were unique to

FIG. 6.

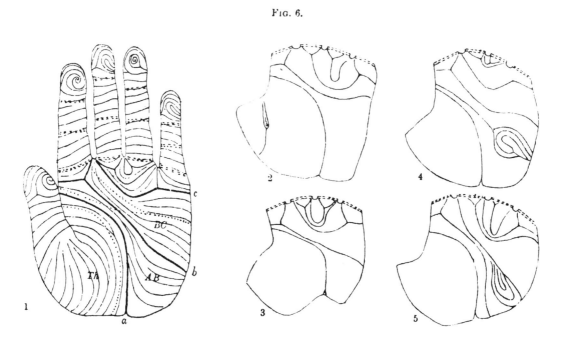

▲ **Ridges of the fingers** and creases of the palm. An illustration from one of Galton's books.

THE MEASURER OF ALL THINGS

Francis Galton, great-grandson of Erasmus Darwin and cousin of Charles, was a remarkable and eccentric figure in Victorian science. After an indifferent education he made a name for himself as a traveller, visiting Africa where he famously used trigonometry to estimate the proportions of a female African Hottentot (Khoikhoi) 'Venus'. On his return he became an important establishment figure in science, serving as secretary of the British Association for the Advancement of Science and helping to set up the journal *The Reader*, which later became *Nature*.

Galton's restless genius expressed itself through a series of inventions, which ranged from a heliostat used for flashing sun signals from onboard ships (commercially manufactured and widely used) to a more eccentric brain cooler (a top hat with a hinged lid) and a supersonic whistle for bothering animals – 'It certainly annoyed some of the lions,' he noted after a trip to London Zoo. He was also credited with the discovery of the anticyclone (high-pressure weather systems) and the invention of the weather map.

These meteorological achievements were indicative of his obsession for measurement, which led him to pioneer biometrics, including the measurement of IQ and other psychological traits. He discovered important statistical principles, such as regression to the mean and correlation coefficients, and used these to advance his 'eugenics' project. Inspired by Darwin's theory of evolution, he became obsessed with artificially evolving the human stock through selective breeding: 'What a galaxy of genius might we not create?' This required a means to select the best specimens, which was where biometrics came in. Galton's interest in fingerprints stemmed from this biometric obsession.

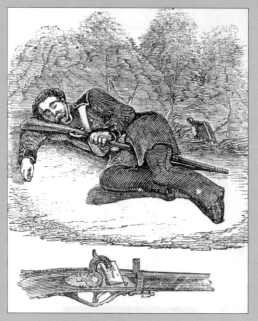

◀ *Galton's classic* **The Art of Travel** *(1855) included handy advice such as how to sleep safely with a loaded gun.*

> *'[Faulds was] a tireless self-promoter who ... After the use of fingerprints was well-established ... did his utmost to write himself back into the picture.'*

GAVAN TREDOUX, 2003

each individual and remained the same throughout life, and devise a system for describing and comparing them. In order to achieve this he needed data, so he got in touch with Herschel. Herschel agreed to provide his records, on the proviso that Galton gave him credit for the work he had done.

Beavan alleges that this constituted a 'secret pact' to cheat Faulds and claims that Galton and Herschel conspired 'to leave Henry Faulds out in the cold'. Tredoux is particularly hard on this idea: '[Beavan's] (ludicrous) allegation of a conspiracy between Galton and Herschel to denigrate Faulds is without foundation and supported only by quotation from Herschel with creative elipsis by Beavan, a process that borders on academic fraud.' It seems more likely that Galton had forgotten about Faulds' letter of 1880, and, as Faulds had produced no other work relating to fingerprinting, there was no reason to pay him any heed.

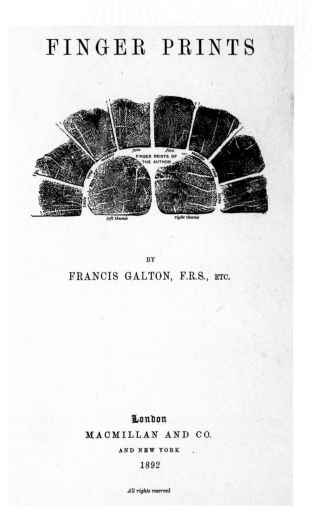

▲ **Marks of genius.** *The title page of Galton's 1892 book on fingerprints featured pictures of his own from both hands.*

In Memory of

*   *

HENRY FAULDS
1843 - 1930
SCOTTISH MEDICAL MISSIONARY TO JAPAN
SURGEON-SUPERINTENDENT OF TSUKIJI HOSPITAL, TOKIO
LEADING PIONEER FROM 1880 IN SCIENCE OF PERSONAL IDENTIFICATION
BY FINGERPRINTS

*   *

SIR WILLIAM J. HERSCHEL
Hooghly, India
Circa 1865

DR HENRY FAULDS
Tokio, Japan
Circa 1877

History records that he was greatly esteemed by all people of Japan, but that after settling in England, with the offer in 1886, of his Fingerprint System freely made to and unwisely declined by New Scotland Yard, he was most
UNJUSTLY

DEBASED, *as a pioneer, from May, 1888, by knight Galton, friend of knight-baronet Henry, Commissioner of The Yard, imbuing him with his opinion of "nothing new that is of value" in his major conception.*

DISOWNED *and* DISDAINED *accordingly from 1901 by Henry.*

DENOUNCED *and* DAMNED *in Parliament on 19th April, 1910 by Sir Winston Churchill, (then as Mr Churchill and as Home Secretary, Executive Head of the Yard, on Henry's Report) as unworthy of any Government recognition.*

DEGRADED *in 1931 by Assistant Commissioner, knight Kendal.*

DISGRACEFULLY *misrepresented in 1935 as a writer unknown in Fingerprint Literature before 1923, as appears from the Third Edition of "Contribution to Bibliography dealing with Crime and Cognate Subjects," compiled by knight Cumming, founder and first Editor of The Police Journal. This catalogue, "after a rigorous revision," purported to be reasonably comprehensive and international, covering in the main a period of fifty years, was published at the expense of the Government.*

DESPISED *and* DISCOUNTED *in 1949 by Assistant Commissioner Howe; and after being*

DEFAMED *in 1953 by Douglas G. Browne and Allan Brock in their book on "Finger Prints," etc., as "presumed" by them to have desired to bring The Yard's System into disrepute, based on information from their friend, ex-Chief Superintendent Frederick E. Cherrill, as well as,*

DEEMED *in their book (which suppresses Herschel's acknowledgment; See Left Panel) to have been anticipated by Herschel in his major conception; and*

DERIDED *finally in 1954 by Cherrill as Scribe of the Yard with his major conception admitted for the first time but mocked and minimised as of no great importance: all with the approval, tacit of Deputy Commissioner Howe and express of Commissioner Nott-Bower, both knights;*

HIS DAUGHTERS
AGNES CAMERON FAULDS and ISABELLA JANE FAULDS
with heartfelt thankfulness
ACKNOWLEDGE, ALL THE TRIBUTES TO THEIR FATHER'S WORK
(beginning in 1905 with those of Otto Schlaginhaufen, Zurich and Tighe Hopkins, London)

NOTING SPECIALLY
I.—SIR WILLIAM JAMES HERSCHEL, BART—1917;

II.—THE REVEREND SIR JOHN C. W. HERSCHEL, BART—1950;
and
III.—THE BRITISH GOVERNMENT—1933-52.

By Grants to them in Express Recognition of their Father's Services to Britain (as, of course, to all the world) from the Bounty Fund at the disposal yearly of the Prime Minister:—

(1) In 1933—Under the Premiership of Mr Ramsay MacDonald of one thousand shillings for each to "roof" the dwelling house of their Father and their home since his death: due to the application on their behalf of the Vicar of Wolstanton, Stoke-on-Trent ;
and
(2) In 1952—Mr Churchill (not then accoladed) being Prime Minister, but whether for each of five thousand shillings or more or less, they are barred from telling here: due to the application on their behalf of their Father's recorder and their nine Supporters, headed by the Executors of the Reverend Sir John C. W. Herschel, Bart.

"He (Faulds's) letter of 1880 announced . . . that he had come to the conclusion, by original and patient experiment, that fingerprints were sufficiently personal in pattern to supply a long-wanted method of scientific identification, which would enable us to fix his crime upon any offender, who left finger-marks behind him and equally well to disprove the suspected identity of an innocent person. (For all of which I gave him and I still do so, the credit due for a conception so different from mine.")

Sir William James Herschel, Bart., in *Nature* of January 18, 1917.

"I examined directly many thousands of living fingers, then passed on to consider impresses on putty, bees-wax, sealing-wax, clay and other substances, taken from my own fingers, those of students under my care, and medical men, native and foreign and out-patients, who might visit the hospital. These were at first very roughly classified and analysed. I am quite sure that at this point the conception [including his major conception] of a wide and general method of identification flashed upon me with such suddenness." Dr Henry Faulds in *Knowledge*, April, 1911, quoted "1938," p. 18.

*   *   *

"I may add that I have not the slightest wish to diminish the credit that may be due to Sir W Herschel." Dr Henry Faulds in *Nature* of November, 22, 1894.

In 1892, Galton's researches bore fruit in his book, *Finger Prints*, and he followed this with a number of other articles and books that made the case for fingerprints as a forensic tool. The authorities began to take note, and in 1894 a parliamentary commission recommended that the police should make use of the technique. It was at this point that Faulds resurfaced, firing off an angry letter to *Nature* in which he effectively accused Herschel of pre-empting his claims for priority. Faulds challenged the judge to produce a copy of a document of 1877 on the topic, as it 'would go far to settle the question of priority'. Herschel promptly produced said document, and pointed out in measured terms that, while he acknowledged that 'Mr. Faulds' letter of 1880 was, what he says it was, the first notice in the public papers … of the value of finger-prints for the purpose of identification … I scarcely think that such short experience as that justified his announcing that the finger-furrows were "for-ever unchanging".' In other words, there was no way that Faulds could have deduced the invariate nature of fingerprints from just two years' work.

Over the next seven years, fingerprinting would become accepted as a useful police tool, thanks to work done by Edward Henry, a police Inspector General in India, in developing the system founded by Galton and Herschel. Faulds continued to agitate for the public recognition he felt had been denied him. In 1905, he wrote a book on the topic, barely mentioning Galton, and was unfavourably reviewed in the *Times Literary Supplement* as 'a man with a grievance'. The critic, Arthur Shadwell, wrote: 'if Mr. Faulds thinks his case will be assisted by writing of this kind he must have a pretty poor opinion of his readers.'

Galton had so far ignored Faulds, but the attack on Herschel prompted Galton to respond. Reviewing Faulds' book in *Nature* he wrote: 'he overstates the value of his own work, belittles that of others, and carps at evidence recently given in criminal cases.' The row rumbled on, and in 1917 Herschel once again asserted that he recognized Faulds' priority when it came to suggesting the forensic use of fingerprints: 'For all of which I gave him and still do so, the credit due for a conception so different from mine.' But Faulds continued to campaign for official recognition, and after his death in 1930 his cause was taken up by Scottish lawyer George Wilton Wilton. More recently, Beavan's book has apparently made Faulds a cause célèbre among Scottish nationalists, despite Tredoux's contention that 'Faulds was in the end little more than an annoyance.'

FREUD

vs

ADLER

FEUDING PARTIES

Sigmund Freud (1856–1939)
– founder of psychoanalysis,
neurologist, psychiatrist, writer
and icon

vs

Alfred Adler (1870–1937)
– founder of school of Individual
Psychology, doctor, international
celebrity, best-selling author and
one-time associate of Freud

DATE

*c.*1910–37

CAUSE OF FEUD

Ideological differences over
primary psychological drives and
role of unconscious, exacerbated
by political and personal
disagreements

In the late 19th century, Sigmund Freud had forged a revolutionary and controversial new approach to the treatment of psychological illness and study of the human psyche. This 'psychoanalytic theory' seemed to offer the potential to transform society, but to many it was dangerous and perverted, with its insistence on the role of childhood sexuality in personality formation. It was central to Freud's project that his ideas should be scientifically validated, and to him this meant two things. Firstly, that large numbers of psychoanalysts should gather large amounts of data that would corroborate his hypotheses by sheer weight of numbers. Secondly, that there should be what he called 'a unified conception' – an agreement between all involved on subject matter, terminology and method.

To pursue the former, Freud encouraged the formation of institutions, beginning in 1902 with the circle of sympathetically minded doctors he invited to his rooms every Wednesday – an informal group that came to be known as the Wednesday Psychological Society. By 1908, this had grown into a much larger association, the Vienna Psychological Society, which in turn spawned similar groups around the world and an umbrella organization, the International Psychoanalytic Association, founded in 1910. But this expansion of the Freudian project inevitably threatened the orthodoxy, or 'unified conception', upon which Freud insisted. Conflict was inevitable, especially given the emotionally and psychologically fraught relationships Freud insisted on developing with his closest colleagues.

TIMELINE

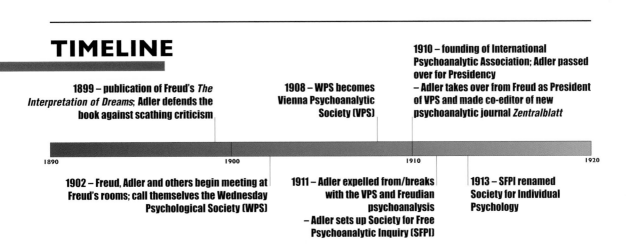

1910 – founding of International Psychoanalytic Association; Adler passed over for Presidency
– Adler takes over from Freud as President of VPS and made co-editor of new psychoanalytic journal *Zentralblatt*

1899 – publication of Freud's *The Interpretation of Dreams*; Adler defends the book against scathing criticism

1908 – WPS becomes Vienna Psychoanalytic Society (VPS)

1890   1900   1910   1920

1902 – Freud, Adler and others begin meeting at Freud's rooms; call themselves the Wednesday Psychological Society (WPS)

1911 – Adler expelled from/breaks with the VPS and Freudian psychoanalysis
– Adler sets up Society for Free Psychoanalytic Inquiry (SFPI)

1913 – SFPI renamed Society for Individual Psychology

Original gangster

Alfred Adler was an ambitious young doctor with an interest in education and psychology who soon made an impact on Freud at their regular meetings: 'He is the only personality there,' Freud wrote to a friend. The overworked older man was able to pass on some of his lucrative caseload to Adler, and approved of his first book, *A Study of Organ Inferiority*, published in 1907. But he was also ambivalent about his young protégé. Ludwig Binswanger recalled visiting a WPS session in 1907, at which Adler and a few others were present. When the meeting was over, Freud turned to Binswanger and asked, 'Well, have you seen these gangsters now?'

Soon theoretical differences between the men began to emerge. Freud stressed ever more dogmatically the central role of sex, especially childhood sexuality, in his theory of psychology. Adler, however, was developing his theory of the 'inferiority complex', in which the primary motivating factor of personality was a struggle for dominance. The genesis of this emphasis, and possibly of his difficult relationship with Freud, fittingly extended back to his own childhood. As he related to the writer Phyllis Bottome, with whom he collaborated on a biography near the end of his life, 'he felt himself put in the shade by a model elder brother … who always seemed to Alfred to be soaring far beyond him in a sphere to which he – for all his efforts – could never attain. Even at the end of his life he had not got wholly over this feeling.' What Bottome failed to mention, perhaps because Adler had not told her, was that the name of this older brother was Sigmund.

Out of the shadows

Freud was increasingly exasperated by Adler's divergence from what he saw as the core concepts of psychoanalysis, and wrote disparagingly of what Adler offered in their

◀ **Superstar of psychoanalysis.** *Alfred Adler arrives in Plymouth, England, in 1936. Adler toured the world giving lectures in the 1930s – his 1927 book* Understanding Human Nature *had been a best-seller and for a time he was the most popular figure in psychology.*

▲ **The Third Psychoanalytic Congress, Weimar, 1911.** *Many of the great names in psychoanalysis are present, with Freud in the centre and Jung to his right. Adler had already been cast out.*

place: 'it consists of three elements … good contributions …, superfluous but admissible translations of analytic facts into a new jargon, and distortions of these facts.' According to Freud, Adler's patients suffered merely from 'untidy character conflicts', but not 'genuine hysterias and big neuroses'; his developmental model was an 'infantile scuffle', which amounted to little more than vulgar clichés, such as 'wanting to be on top' and 'covering one's rear'. Freud was also losing patience with the man himself, writing to Carl Jung in November 1910 that: 'My mood is not good because of annoyances with Adler … who [is] hard to get along with.'

Adler, meanwhile, was nursing his own grudges. He felt restricted and unappreciated, and matters came to a head in 1910 when Freud, eager that the new International Psychoanalytic Association not be seen simply as a vehicle for Viennese Jews (of which Adler was one), chose his new golden boy Jung to become its president and to take over as editor of psychoanalysis' flagship journal. Adler was furious. Freud tried to placate him by resigning as president of the VPS in his favour and by founding a new journal, *Zentralblatt*, for him to co-edit; but it wasn't enough. Each man had a different take on what happened next, but most accounts agree that there was a showdown at which Adler famously lamented, 'Why should I always work in your shadow?' According to Freud, Adler's 'striving for a place in the sun' necessitated that he be told to leave. Adler's account is that Freud begged him to stay, but that 'it was no pleasure to stand in his shadow', particularly when that shadow blackened his own theories by association with Freud's fixation with childhood sexuality. This version is partly backed up by a letter he wrote soon after: 'Since I have no desire to carry on such a personal fight with my former teacher, I herewith tender my resignation.'

PSYCHOANALYST, HEAL THYSELF

Freud's self-analysis was a major step in his development of psychoanalytic theory, yet he seemed remarkably ill-equipped to understand or cope with the intensity and destructiveness of the close relationships and bitter break-ups that marked his professional life. Time and again he repeated the same pattern of intense infatuation with someone who became an important colleague and confidant, only to fall out with them amid a welter of accusations of paranoia, pettiness and perversion. Adler and Jung are only the better-known names. Others include Albert Moll, Wilhelm Fliess and Josef Breuer. Ostensibly, the reasons for the breaks included spats over priority, professional and personal timidity in the face of Freud's radical and disturbing views on childhood sexuality, and deviation from Freud's increasingly dogmatic conception of psychoanalysis. Fittingly, however, psychodynamic currents seethed below the surface.

A core concept of Freud's was that in childhood everyone develops relationship templates that in later life are 'transferred' onto other people and situations. As Freud scholar Douglas Davis comments, 'He himself created a profound transferential wake, in which most of those who became his associates found themselves awash.' In childhood, Freud had felt resentment and bitterness towards a brother who later died, towards cousins with whom he had played out an infantile love triangle and inevitably towards his own father. He would transfer these complexes onto figures such as Fliess and Jung, competing with them for the attentions of female patients, setting himself up as a father figure for whose affection the children had to compete, and wishing failure and doom to those who challenged him. 'You go about … reducing everyone to the level of sons and daughters who blushingly admit the existence of their faults. Meanwhile you remain on top as the father, sitting pretty,' Jung wrote to him around the time of their break.

'Snatching and pilfering and all the scholar's shabby machinations'

ADLER ON FREUD, 1913

Putting the boot in

Adler departed from the VPS and set up his own organization, the Society for Free Psychoanalytic Inquiry (SFPI), a name that Freud and his adherents took to be a slight. Later, cutting all ties with psychoanalysis, Adler renamed his own school Individual Psychology. The two men quickly became embittered enemies. In 1913, Adler wrote

THE VIENNA SCENE

The crucible of psychoanalysis, fin-de-siècle Vienna, was a remarkable moment in place and time. Freud's family moved there when he was four, and it was in Vienna that he was educated and spent the formative years of his medical and neurological training. The city was the capital of Austro–Hungary, an empire in terminal decline but of glittering cultural and intellectual brilliance. As the centre of government and commerce, Vienna drew on every corner of the ethnically diverse empire to become one of the world's first great multicultural cities: it was peopled by Teutons, Hungarians, Czechs, Slovenians, Croats, Serbians, Ruthenians, Bosnians and Jews.

Vienna was a city of remarkable potential and oppressive tension; fertile ground for all the diverse forms of modernism. It was home to the Secessionist art movement of Klimt and Schiele; the philosophical movements of positivism and sensualism, including scholars such as Wittgenstein and Mach; new literature, with writers such as Schnitzler; and a font of scientific and medical research – most notably Freud's project for a scientific psychology. Alongside these progressive forces were violent reactionary ones: the repressive forces of bourgeois culture and ethnic tensions that found expression in virulent anti-Semitism (it was in Vienna that Hitler formed his poisonous ideology). Freud's psychoanalysis, his difficult and combative personality, and the conflicts and feuds that resulted – all these were the product of their place and time.

▲ *Family ties. The extended family of Jacob Freud c.1878, with young Sigmund third from the left in the back row: a picture of bourgeois, assimilated respectability, but what psychodynamic currents seethed beneath the surface?*

to a friend on the subject of Freud and his gang: 'I see only … busy snatching and pilfering and all the scholar's shabby machinations.' Later in life he would fly into a rage if it were suggested that he had ever been a pupil of Freud's (despite considerable evidence that this was how he saw himself at the time), and referred to Freud's theory of psychoanalysis as 'filth' and 'faecal matter'.

Freud was hardly less damning, writing to his disciple Lou Andreas-Salomé in July 1914: '[Adler's] letter shows his specific venomousness, is very characteristic of him … he is a disgusting person.' His antipathy outlasted Adler, as evidenced by a shocking and bitterly sarcastic letter he wrote to Arnold Zweig shortly after the younger man's lonely death in Scotland in 1937: 'I don't understand your sympathy for Adler. For a Jew boy out of a Viennese suburb a death in Aberdeen is an unheard of career in itself and a proof of how far he had got on. The world really rewarded him richly for his services in having contradicted psychoanalysis.'

FREUD
vs
JUNG

FEUDING PARTIES
Sigmund Freud (1856–1939)
– founder of psychoanalysis,
neurologist, psychiatrist,
writer and icon
vs
Carl Gustav Jung (1875–1961)
– founder of depth or Jungian
psychology, psychiatrist
and mystic

DATE
1912–39

CAUSE OF FEUD
Ideological, theoretical and
personal differences

O f all Freud's protégés, Jung was arguably the one in whom he invested the greatest hopes, which must have made what was perceived as his betrayal all the more wounding. A once adoring and fervent relationship degenerated into murky accusations of anti-Semitism.

'Altogether remarkable'

At the start it had all been wine and roses. Jung, a Swiss psychiatrist practising at the prestigious Burghöltzi, an asylum and clinic in Zurich, became an avid fan of Freud in 1906. Freudian psychoanalysis influenced Jung's first book, a study on schizophrenia called *The Psychology of Dementia Praecox*. A correspondence began and in 1907 the two met for the first time, talking for 13 hours. 'I found him extremely intelligent, shrewd, and altogether remarkable,' the Swiss wrote. The admiration was mutual; Freud was impressed with his energy and forceful personality, his gifts for research and his insights, and not least by the opportunity that Jung represented for psychoanalysis to move out of its Viennese 'ghetto'.

As their infatuation deepened, Jung went so far as to admit, in a letter of 1907: 'My veneration for you has something of the character of a "religious" crush. Though it does not really bother me, I still feel it is disgusting and ridiculous because of its undeniable erotic undertone.' Freud, too, experienced powerful emotions. Superficially, he cast himself in a paternal role, with Jung as his heir, but on at least two occasions he fainted in Jung's presence, having to be carried to his couch.

Revolt of the egg

In 1909, Jung joined Freud on an important tour of the United States, and in 1910 he was officially anointed as the international face of psychoanalysis when Freud made him president of the International Psychoanalytic Association. Even before the two men had met, however, Jung had been aware of potentially divisive differences in their approaches to psychoanalysis. Jung had always been fascinated by the spiritual and mystical – areas that Freud regarded with mistrust – and was

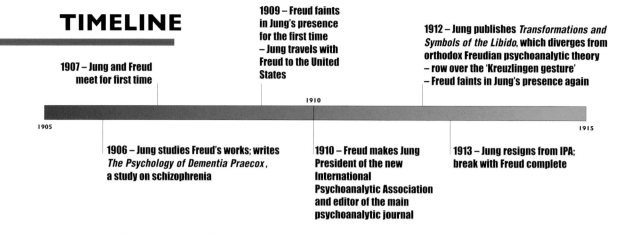

TIMELINE

1907 – Jung and Freud meet for first time

1909 – Freud faints in Jung's presence for the first time – Jung travels with Freud to the United States

1912 – Jung publishes *Transformations and Symbols of the Libido*, **which diverges from orthodox Freudian psychoanalytic theory – row over the 'Kreuzlingen gesture' – Freud faints in Jung's presence again**

1910

1905

1915

1906 – Jung studies Freud's works; writes *The Psychology of Dementia Praecox*, **a study on schizophrenia**

1910 – Freud makes Jung President of the new International Psychoanalytic Association and editor of the main psychoanalytic journal

1913 – Jung resigns from IPA; break with Freud complete

cautious about the older man's emphasis on sexuality and, particularly, infantile sexuality. 'Every form of addiction is bad,' he later wrote, 'no matter whether the narcotic be alcohol or morphine or idealism.'

Like Adler before him, Jung developed his own theories, challenging Freud's attempt to impose orthodoxy and straining their relationship. '[It's] a risky business for the egg to try to be cleverer than the hen,' Jung admitted in 1911; but the next year he produced his most important book to date, *Transformations and Symbols of the Libido*. 'The whole thing came upon me like a landslide … it was the explosion of all those psychic contents that could find no room, no breathing space, in the constricting atmosphere of Freudian psychology and its narrow outlook,' he later explained.

The seeds of discord were now sown. Freud was upset at what he saw as a challenge to his carefully constructed edifice, and his concern turned to hostility when Jung returned to America in 1912 and lectured on his new theories as well as Freud's. On Jung's return the two swapped increasingly bad-tempered letters. Quoting Nietzsche, Jung told Freud: 'One repays a teacher badly if one remains only a pupil.' In another letter he wrote: 'If ever you should rid yourself entirely of your complexes and stop playing the father to your sons, and instead of aiming continually at their weak spots took a good look at your own for a change, then I will mend my ways and at one stroke uproot the vice of being in two minds about you.'

▶ **Hello America.** *Freud and Jung, seated on the left and right, on their tour of America in 1909, here pictured at the Clark University Psychology Conference.*

By this time the two had fallen out over the notorious 'Kreuzlingen gesture', a petty dispute that illustrated how fraught relations had become. Freud had visited a colleague in Kreuzlingen in Switzerland, and was upset that Jung had not bothered to come and meet him. Jung, meanwhile, said that he was hurt that Freud had not informed him he was coming to the country. In a tense showdown, Freud made Jung admit that he had been aware of the visit. There was a difficult lunch, and during a discussion in the afternoon in which Jung alluded to the psychodynamics of conflict within the psychoanalytic movement, Freud fainted again. 'Just like a woman. Confront her with a disagreeable truth: she faints,' sneered Jung.

Tension mounted, until Jung snapped and shot off a letter that began ominously, 'May I say a few words to you in earnest?' He laid into Freud's character, his neuroses, his self-awareness and even his abilities as a therapist. Freud was bitterly angry; he offered a restrained reaction at first, simply suggesting that they break off their personal relationship, but by the following year he was describing Jung as 'outrageously insolent' and complaining that he was a 'florid fool and brutal fellow'.

▼ **Freud's study/consulting room,** complete with couch, recreated in Hampstead, London, where he lived for the last year of his life after being forced to flee the Nazis.

> *'For sheer obsequiousness nobody dares to pluck the prophet by the beard.'*
>
> JUNG TO FREUD, 1912

Dark night of the soul

Jung struggled on as president of the IPA until opposition from Freudians convinced him to resign in April 1913; by 1914 his professional break with Freud was complete. Over the next six years he would struggle to deal with the psychic fallout from the break-up, while at the same time exploring and developing his own brand of psychoanalysis, which came to be called Analytical Psychology. It differed from Freudianism in its insistence on the importance of the mystical and spiritual, and in Jung's theory of the 'collective unconscious', a kind of group mind shared by all humanity, inhabited by archetypes. In his 1939 *Moses and Monotheism*, which can be seen as a riposte to Jung's emphasis on myth and mysticism, Freud dismissed the concept: 'I do not think that much is to be gained by introducing the concept of a "collective" unconscious – the content of the unconscious is collective anyhow, a general possession of mankind.'

But the split between the two men was more than academic. Part of Jung's attraction for Freud was his non-Jewish, Aryan nature. As the Nazis took control of Germany and later Austria in the 1930s, the Jewish dimension of psychoanalysis became the object of attack, and Jung's role in the growing anti-Semitism was controversial. Although on the one hand he maintained links with Jewish colleagues in the face of Nazi disapproval, on the other he was linked with aspects of the Nazification of psychiatry, seemed ambivalent about Hitler, and wrote provocatively of the split between his school and those of Freud and Adler in decidedly anti-Semitic terms: 'I can understand very well that Freud's and Adler's reduction of everything psychic to primitive sexual wishes and power-drives has something about it that is beneficial and satisfying to the Jew, because it is a form of simplification.'

Jung's adherents and subsequent apologists have marshaled extensive evidence to rebut claims of anti-Semitism, but passages like the one above lend weight to Freud's bitter accusation that Jung was guilty of 'lies, brutality and anti-Semitic condescension to me'. In 1924, Freud went so far as to call Jung an 'evil fellow'. Jung was more respectful of his former teacher, acknowledging his contribution while pointing out his shortcomings.

SABIN

vs

SALK

FEUDING PARTIES
Albert Sabin (1906–93)
– microbiologist
vs
Jonas Salk (1914–95)
– microbiologist

DATE
1950s–90s

CAUSE OF FEUD
The search for a safe polio
vaccine

By the early years of the 20th century, public-health advances had, perversely, made children more vulnerable to infection by the poliovirus, a micro-organism that can cause the paralyzing and sometimes fatal condition poliomyelitis. In the US, fear of the condition lent strength to the fundraising efforts of the National Foundation for Infantile Paralysis (NFIP, later renamed the March of Dimes); it became the best-funded health organization in the country and its president, Basil O'Connor, became one of the most powerful men in US science.

The NFIP funded the aggressive pursuit of a vaccine, and in 1935 trials were attempted of both 'killed virus vaccine' (where batches of virus are deactivated by exposure to toxic agents such as formaldehyde) and 'attenuated/live virus vaccine' (which uses living but weak (attenuated) strains of virus that can stimulate an immune response but will not cause full-blown disease – at least, not in most individuals). Not only did the trials fail, but scientists suspected the faulty vaccines might have caused paralysis and death in some of the children exposed to the vaccines.

By the 1940s, a number of important advances, including the successful culturing of poliovirus in tissue culture by John Enders (1897–1985) and colleagues, meant that a safe and successful vaccine was now within reach. Whoever secured the formidable backing of O'Connor and the NFIP might well achieve this feat and become the greatest medical hero of the age.

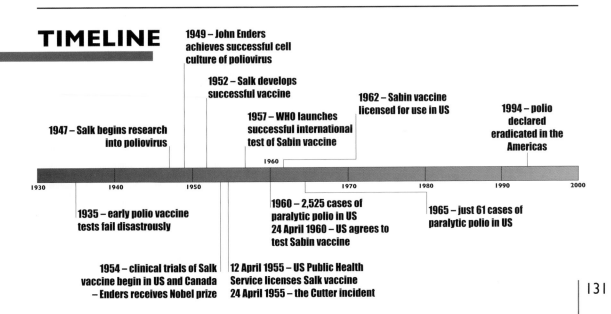

TIMELINE

1949 – John Enders achieves successful cell culture of poliovirus

1952 – Salk develops successful vaccine

1962 – Sabin vaccine licensed for use in US

1994 – polio declared eradicated in the Americas

1957 – WHO launches successful international test of Sabin vaccine

1947 – Salk begins research into poliovirus

1960

1930   1940   1950   1960   1970   1980   1990   2000

1935 – early polio vaccine tests fail disastrously

1960 – 2,525 cases of paralytic polio in US
24 April 1960 – US agrees to test Sabin vaccine

1965 – just 61 cases of paralytic polio in US

1954 – clinical trials of Salk vaccine begin in US and Canada – Enders receives Nobel prize

12 April 1955 – US Public Health Service licenses Salk vaccine
24 April 1955 – the Cutter incident

'The dog's work'

To the chagrin of most of the biomedical establishment in the US, O'Connor picked young microbiologist Jonas Salk, who was able to build on the discoveries of others to create, test and prepare for human trials a vaccine made by inactivating the virus with formaldehyde (aka formalin). Few scientists have divided opinion as much as Salk. According to David Oshinsky, author of *Polio: An American Story*, Salk was prepared to do 'the dog's work that his betters refused to do'. One of Salk's biographers, Jeffrey Kluger, described him as 'a tectonic force in scientific history'. Through leadership and intense hard work, Salk and his huge team produced a working vaccine in 1952. In April 1955 the vaccine, which required a course of injections, was rolled out across the country.

Overnight, Salk became an American hero, but in soaking up the plaudits he managed to alienate almost everybody. 'Once the goal was reached,' writes Oshinsky, 'the group

▼ **The scourge of polio.** *Young polio patients with their leg braces and wheelchairs, reading mail from home while staying at a treatment centre sponsored by President Roosevelt, 1938.*

'[Salk] never had an original idea in his life.'

ALBERT SABIN, 1990

◀ **Egyptian relief** *of ancient polio victim. In fact for most of history poor sanitation meant infants were exposed to the virus early, while still protected by maternal antibodies passed on in breast milk. Advances in public health meant that although children were less likely to be exposed to the virus, if such an exposure did happen it was much less likely to coincide with this period of breastfeeding and maternal protection.*

would split apart amidst charges that Salk had not appreciated, much less acknowledged, the collaborative nature of his success.' Instead he was accused of grabbing the glory for himself: 'One of his greatest gifts was a knack for putting himself forward in a manner that made him seem genuinely indifferent to his fame, a reluctant celebrity.'

Even as public recognition was showered upon him, Salk was pointedly snubbed by his peers. In 1954, a Nobel Prize was awarded for poliovirus research, but it went to Enders and his colleagues. While many others involved in the race for the vaccine were elected to the prestigious National Academy of Sciences, Salk was not. His peers would not buy into what Sabin biographer Angela Matysiak calls 'the myth of Jonas Salk'. According to leading vaccinologist Paul Offit, 'many of his colleagues dismissed Salk as a lightweight or a fake.'

'No one was more critical, more mean spirited, or more persistent in his attacks than Albert Sabin,' says Offit. Sabin was pursuing an alternative form of vaccine, using live, attenuated virus, and conducted a forceful campaign to promote his approach over Salk's. He accused O'Connor of partiality, and insisted, 'A killed-virus vaccine for poliomyelitis must be safe without qualifications. If it is admitted that it can be made safer, then it is not sufficiently safe.' He copied his criticisms to Salk with a vaguely threatening covering letter: 'Dear Jonas, This is for your information – so that you'll know what I am saying behind your back. This incidentally is also the opinion of many others whose judgement you respect. "Love and Kisses" are being saved up. Albert.' On one occasion, Salk recalled, Sabin even telephoned him to accuse him of 'misleading the public'. 'I was flabbergasted,' admitted Salk.

 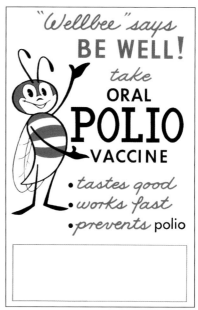

▲ **US national health drive posters:** Tom Little's blunt but effective Pulitzer Prize-winning cartoon, for Salk's vaccine (L); public health mascot Wellbee, pushing Sabin's oral vaccine (R).

The Cutter Incident

The battle between Sabin and Salk began, at times, to resemble the AC/DC War between Westinghouse and Edison (*see* pages 196–203), with both men trumpeting the safety of their vaccine and highlighting the danger of their opponent's. Sabin struck an early blow with the Cutter incident of April 1955: a number of children who had been injected with Salk's vaccine developed polio and 11 died. Officially, the tragedy was caused by improper preparation of a batch at the Cutter Laboratories in California, but suspicions still linger as to whether the Salk formaldehyde process might be inherently unsafe.

Despite the protestations of Sabin, who wrote to Basil O'Connor to complain that 'When such a tragedy occurs you do not continue operations as usual', the incident was explained away and the Salk project continued. Further controversy erupted in the late 1950s when many in the US medical establishment resisted the introduction of the Sabin vaccine. His version could be administered orally via a sugar cube, threatening one of the few guaranteed revenue streams for doctors, who were paid to inject the Salk vaccine. The refusal to adopt the cheaper version meant whole swathes of society could not afford protection from polio.

VACCINATION CONTROVERSIES

Suspicion of vaccination dates back at least as far as the experiments of Edward Jenner, who in 1796 inoculated eight-year-old James Phipps with cowpox, and later tried to infect him with smallpox to see if the strategy had worked. Many have accused Jenner of acting unethically – this is probably unfair but there was widespread resistance to vaccination at the time and throughout the 19th century. Anti-vaccination campaigner George Gibbs blasted the idea of inoculation with 'a loathsome virus from the blood of a diseased brute,' and lobbied against attempts to make vaccination compulsory, complaining of 'medical spies forcing their way into the family circle'. In recent years, a similar nexus of fear and suspicion has cast a shadow over the combined vaccine for measles, mumps and rubella (MMR) (see pages 136–139). Despite being discredited, spurious claims made by Dr Andrew Wakefield are still being touted as proof of a link between the MMR injection and autism. Public figures such as actor Jim Carrey and model Jenny McCarthy continue to recycle baseless lies about vaccines and autism, and there is a subculture of suspicion towards vaccines in general, both in the developed and developing worlds. In Nigeria, UN-sponsored polio-vaccination programmes have been accused of being part of a pan-global conspiracy to target Muslims. As a result, serious outbreaks of the disease have claimed lives. In the developed world, incidence of measles has risen by 2,000 per cent in the last seven years as a direct result of fears about the MMR vaccine.

'Safe without qualifications'

Eventually, Sabin's live-virus vaccine, which had many advantages over Salk's but also some disadvantages, supplanted the killed-virus version in most of the world, helping to nearly eradicate polio across the globe. Sabin would later boast, 'I developed *the* vaccine, not *a* vaccine.' He maintained his hostility towards Salk until near the end of his life; in an interview at the age of 84 he called Salk 'a kitchen chemist'. According to Kluger, Salk shrugged off the attacks: 'Stupidity always made him angry; malevolent stupidity made him angrier still. He wouldn't show it; he never did. You couldn't run the kind of lab he ran and conduct the kind of research he conducted and allow yourself the luxury of pique.' In a twist that Salk might have appreciated, the US now regards his vaccine as the only one guaranteed to be 'safe without qualifications', and it is Salk's vaccine that children in the UK now receive.

SCIENCE AND THE MEDIA

Science is indisputably important to everyone, so everyone should know about science. Yet few people are directly engaged in science and science education is cursory and waning. It thus falls to the media to present science and, in doing so, to shape public attitudes about safety, ethics and funding. These attitudes have direct consequences for scientists, in terms of which fields are funded by government, charities and industry, how science is regulated and which fields are restricted or even banned. The media's influence also has direct consequences for the public, in areas such as consumer choice and moral panics – for example, MMR vaccine uptake or mobile-phone mast siting. The ability of the public to assess risks and use scientific evidence to make decisions about policy or purchase has become dependent on media presentation. So, how good a job is the media doing?

Sensationalizing science

In 2006, a report by the Social Market Foundation, an independent UK research group, concluded that the media was guilty of sensationalizing science through irresponsible and poor reporting. The *Guardian* science columnist and persistent critic of the media's record on science Ben Goldacre puts it rather more bluntly: 'Why is science in the media so often pointless, simplistic, boring, or just plain wrong? ... It is my hypothesis that in their choice of stories, and the way they cover them, the media create a parody of science, for their own means. They then attack this parody as if they were critiquing science.'

Goldacre and the SMF report effectively raise the spectre of British scientist and novelist C.P. Snow to account for the worsening failures of science journalism. In 1959, Snow delivered a lecture warning that the two 'cultures' of science and the humanities were drawing apart. Where before many people had a solid background in both worlds, he argued, changes in education and society were leading to increasing polarization of the two worlds and ignorance of one another. In particular, he accused those in the humanities culture of wilful ignorance and chauvinism against science. While many, then and now, argued that Snow's thesis was mistaken or overstated, the evidence appears to show that he was prescient to a degree, and that many of his worst fears have become realized.

Scare stories

One of the biggest science stories of the last decade has been the MMR vaccine controversy, in which Dr Andrew Wakefield and a tiny number of other researchers suggested a link between the MMR (measles, mumps and rubella) vaccination and autism. Very few of these anti-vaccine campaigners have published peer-reviewed studies to back up their claims and those that have been published have been conclusively shown to be too small to form any conclusions, of sloppy design and, in some cases, even the result of misconduct and fraud. By contrast, rigorous research involving huge sample sizes has consistently failed to find any connection whatsoever between MMR and autism. Yet the tone and content of media reporting over a period of several years failed to reflect the true status of the MMR claims and contributed directly to a massive downturn in MMR-vaccine take-up, which in turn has led to increases in the incidence of the diseases, measles in particular, and the deaths of some children.

More recent vaccine-related stories in the UK relate to a cervical cancer vaccine injection, which has prompted wildly opposing reactions by the same newspaper. The *Daily Mail* newspaper, which has consistently talked up vaccine concerns, ran headlines in its UK edition that questioned the safety of the new jab – 'How safe is the cervical cancer jab? Five teenagers reveal their alarming stories' – while in its Irish edition it blasted the government for withholding the jab – 'Join the *Irish Daily Mail*'s cervical cancer vaccination campaign today'.

Dr Andrew Wakefield arrives at a General Medical Council hearing in London in 2007. Wakefield was accused of misconduct over the way he carried out research into alleged links between the MMR jab and autism.

A different type of scare story is one where headlines are generated on the thinnest of evidence. In April 2009, there was a slew of headlines claiming that new research showed that the internet application Twitter makes people 'immoral'. The research cited related blood flow in the brain to experience of types of compassion but never actually mentioned the internet, let alone Twitter. As one of its authors, Professor Antonio Damasio, commented: 'As you can see if you read our study, we made no connection whatsoever with Twitter. Some writers did make that connection but it is not ours. There is no mention whatsoever of Twitter or of any social network in our study. We have nothing whatsoever to say about them.'

Frankenstein's monster

Goldacre contrasts 'scare stories' with two other categories of story: 'wacky' and 'breakthrough' stories. All three types tend to misrepresent the actual process of science and the generally careful language of scientific papers, and act to produce and reinforce a specific – albeit often contradictory image – of science that harks back to the 1950s. Echoing Snow, Goldacre explains that this image actually represents 'the humanities graduate journalists' parody of science, for which we now have all the ingredients: science is about groundless, incomprehensible, didactic truth statements from scientists, who themselves are socially powerful, arbitrary, unelected authority

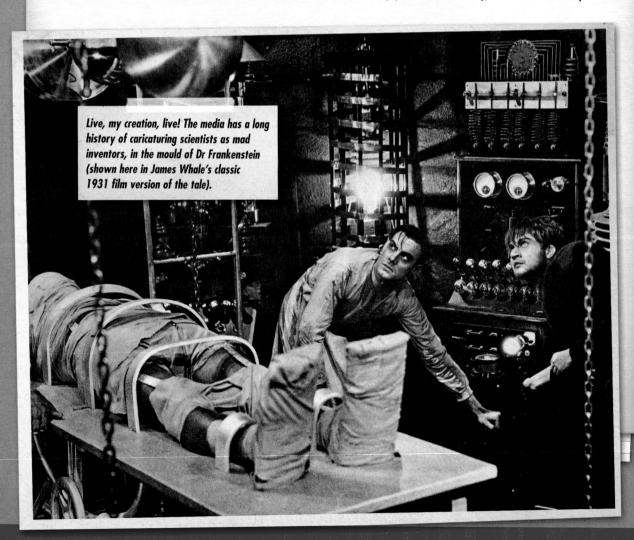

Live, my creation, live! The media has a long history of caricaturing scientists as mad inventors, in the mould of Dr Frankenstein (shown here in James Whale's classic 1931 film version of the tale).

figures. They are detached from reality: they do work that is either wacky, or dangerous, but either way, everything in science is tenuous, contradictory and, most ridiculously, "hard to understand".'

This parody in turn dates back to the Romantic era, when literary figures such as Blake, Shelley and Wordsworth reacted against the Scientific Revolution by caricaturing scientists as soulless automatons, culminating in Mary Shelley's *Frankenstein*. Ever since, the media have struggled to accommodate the messy realities of science and instead have tried to shoehorn it into neat narratives.

One result is that scientists themselves have long had an uneasy relationship with the media. Exposure can turn scientists into superstars – Stephen Hawking and Richard Dawkins are latterday examples; Albert Einstein and Wernher von Braun were the media darlings of their day. Von Braun's unlikely journey – from Nazi bombmaker to avuncular Disney expert – epitomizes the power of the media to shape the careers of scientists. During World War II he built V2 rockets for the Nazis; he was an SS officer and implicated in the use of slave labour. After the war, he oversaw the American rocket programme, and in 1955 Walt Disney recruited him to star in a series of films about the space race that made him the most popular scientist in America.

The media has also had a tendency to project chosen subjects as 'science heroes'. Jonas Salk, who led the killed-virus polio-vaccine project in the 1940s and '50s (*see* pages 130–135), was a prime example. Apparently happy to accept sole glory for what had been a collaborative effort, Salk was transformed almost overnight into a paragon of science, a 'Respected Authority' on whose radio addresses the nation would hang. The scientist as hero still exists in today's media, but scepticism and suspicion have long since transformed the public's relationship to its establishment and authority figures, with the result that modern scientific heroes tend to be figures like Dr Andrew Wakefield, portrayed as lonely figures struggling against repressive institutional forces.

Werner von Braun. Despite having used slave labour to build Nazi bombs, von Braun later became the star of a series of Disney films about the space race that captured an audience of 42 million.

FRANKLIN
vs
WILKINS

FEUDING PARTIES
Rosalind Franklin (1920–58)
– crystallographer
vs
Maurice Wilkins (1916–2004)
– molecular biologist;
James D. Watson (born 1928)
– molecular biologist

DATE
1951–53

CAUSE OF FEUD
Personality clash during the race
to unlock the secret of DNA

The tale of the race to discover the structure of DNA is one of the most exciting in modern science, and also one of the most popular thanks to *The Double Helix*, the best-selling account by James Watson. In late February 1953, Watson and his colleague Francis Crick (1916–2004) worked out that DNA was a double helix, made up of two intertwining chains of molecules like a spiral ladder. Each chain has a backbone of linked sugar molecules, which form the sides of the ladder, and projecting from these are molecules called nucleotide bases, which make up the rungs. The bases bond with their opposite numbers in a specific and ordered way, and it is this which allows DNA to function as the template for genetic information, and thus as the blueprint for all forms of life.

Cracking the DNA enigma was one of the greatest breakthroughs in science, and Frances Crick and James Watson won a Nobel prize for it, along with Maurice Wilkins, who had done important work on imaging the DNA molecule through a process known as X-ray crystallography. However, there was a fourth person, Rosalind Franklin, who had also made a major contribution to this imaging work, without whose labours Watson and Crick would not have made their breakthrough ahead of the competition – competition that included Franklin herself. Why did she miss out on the discovery of DNA, and why did she not get the credit she deserved? Sadly, the simplest explanation was that she and Maurice Wilkins did not get on, and that a simple workplace feud cheated Franklin of a place in history.

▶ **Young Turks.** *James Watson (L) and Francis Crick (R) at Cambridge in 1953, with their model of the molecular structure of DNA.*

Nobody's assistant

Rosalind Franklin was a determined character; a young woman who chose a scientific career at a time when few women even went to university. According to her biographer, Anne Sayre: 'possibly the strongest consequence of brief and early ill-health was that it set firmly her pattern of reacting to frustration with passionate indignation, of fighting every inch of the way rather than submitting for a moment.' Her colleague and friend Aaron Klug admits that, 'She was single-minded and uncompromising in her work, so that she sometimes was bound to provoke exasperation among her colleagues,' yet points out that 'she was not austere, she had a sense of fun.'

Franklin's spiky manner was to cause problems when she was recruited to King's College, London, by the head of the biophysics laboratory, John Randall, to bring her expertise in the art of X-ray crystallography to bear on the laboratory's nascent investigation of DNA. Randall, who many suspect of being the true villain of the piece, specifically told her: 'as far as the experimental X-ray effort is concerned there will be at the moment only yourself and [Raymond] Gosling.' Gosling had been the assistant of Maurice Wilkins, a young biophysicist who had just started to sink his own teeth into the DNA problem, but who returned from holiday to find Franklin had been given his task, his graduate student and even the high-grade samples of DNA he had personally secured.

Wilkins, unaware of Randall's assurances to Franklin, assumed at the very least that she was there to collaborate with him; he may even have believed that she was supposed to report

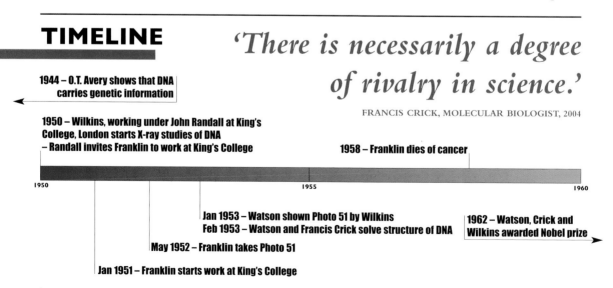

TIMELINE

'There is necessarily a degree of rivalry in science.'

FRANCIS CRICK, MOLECULAR BIOLOGIST, 2004

1944 – O.T. Avery shows that DNA carries genetic information

1950 – Wilkins, working under John Randall at King's College, London starts X-ray studies of DNA
– Randall invites Franklin to work at King's College

1958 – Franklin dies of cancer

1950   1955   1960

Jan 1953 – Watson shown Photo 51 by Wilkins
Feb 1953 – Watson and Francis Crick solve structure of DNA

1962 – Watson, Crick and Wilkins awarded Nobel prize

May 1952 – Franklin takes Photo 51

Jan 1951 – Franklin starts work at King's College

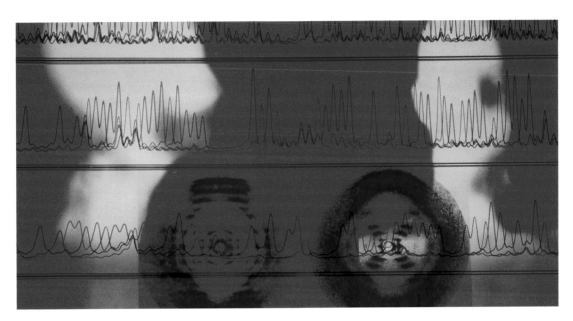

▲ **Pieces of the puzzle.** *A pair of X-ray diffraction patterns revealing the helical nature of the DNA molecule overlaid on a modern DNA sequence trace showing the sequence of bases that make up the DNA code.*

to him. Franklin, wary of being forced into a subordinate role by sexist assumptions, bristled. '[She] claimed that she had been given DNA for her own problem and would not think of herself as Maurice's assistant,' explains Watson in *The Double Helix*.

A little communication and compromise might have soothed over any incipient misunderstanding, but sadly Wilkins' personality was diametrically opposed to Franklin's. In his landmark account of the birth of molecular biology, *The Eighth Day of Creation*, Horace Freeland Judson writes: 'Some who worked with [Wilkins] remarked that he could respond to vigorous disagreement only by turning aside.' The mild-mannered Wilkins simply could not cope with the confrontational style of Franklin. 'Almost from the moment she arrived in Maurice's lab, they began to upset each other,' notes Watson.

'Wilkins continued to work on DNA, but from much poorer-quality samples, and from this point there was in effect no collaboration and very little communication between him and Franklin,' comments historian of biology Robert Olby. 'Wilkins was his own worst enemy. Had he stood up to Randall and Franklin, refused to leave the bench, and demanded effective collaboration, the leadership in the pursuit of the structure of DNA might not have been snatched from the group at King's.'

▲ **Nobel Prize winners, 1962.** *Wilkins is far left, Crick is third from left and Watson is second from right. Between Crick and Watson is the author John Steinbeck.*

'The best place for a feminist'

She may not have been to blame, but Franklin now found herself on the wrong side. Wilkins was friends with Crick and Watson, who were undertaking their own research at Cambridge University, and Watson detailed the sourness that now coloured the men's opinions of Franklin. In *The Double Helix,* Watson portrays an obstructive harridan – 'not a trace of warmth or frivolity' – without the imagination to embrace the creative approach to science needed in the race to unravel DNA. At several points he discusses her in decidedly sexist terms – 'she was not unattractive and might have been quite stunning had she taken even a mild interest in clothes' – and concludes, 'Clearly [Franklin] had to go or be put in her place … the best place for a feminist was in another person's lab.'

Later in the book Watson makes it clear that he has wronged Franklin: 'it became apparent to us that [her] difficulties with Maurice … were connected with her understandable need for being equal to the people she worked with.' In the epilogue he has only warm words for her but the damage had been done. A vital piece of evidence in the hunt for the solution to the DNA problem was Photo 51, an X-ray picture Franklin had obtained of DNA crystals that clearly demonstrated the helical nature of

> ## 'Some people are over-competitive – they have what's known as "Nobel-itis"'
> FRANCIS CRICK, MOLECULAR BIOLOGIST, 2004

the molecule. She herself had ignored it for the best part of eight months, her feud with Wilkins and the others blinding her to its importance, and by the time she started to appreciate its significance, in January 1953, Wilkins had already shown it to Watson.

By the end of February, using this and other clues from Franklin's work, Watson and Crick put the pieces together and solved the puzzle. It is generally acknowledged that Franklin and Wilkins, separately, were not far behind them – Crick himself later estimated that Franklin was just three weeks to three months away from solving the structure of DNA. If they had overcome their personal differences perhaps they would have got there much faster, but in the words of H.B. Fell, who worked in the King's College lab with both of them: 'They were both, I think, rather difficult people.'

THE MISSING NOBEL

One of the most common perceptions of the DNA 'controversy' is that Franklin was denied her fair share of the 1962 Nobel prize, awarded to Crick, Watson and her erstwhile enemy Wilkins. This snub is generally regarded as the ultimate insult, added to the injuries inflicted on her career and reputation by the antipathy of Wilkins and Watson's unflattering account. Indeed, the perception that Franklin was denied recognition that was due to her has made her a feminist icon and a symbol of women's struggle to overcome gender barriers in science.

Much of this is probably fair: for too many years Franklin was largely written out of the legend of the double helix, while some of the difficulties she endured during her short career testified to unacceptable levels of sexism. What is often overlooked, however, is that the central charge laid by Franklin advocates – that she was unfairly excluded from Nobel glory – is baseless, because her tragically early death meant she could never have been awarded it. As Francis Crick explained, 'It's not possible to have it posthumously … Moreover there would have been a problem for the electors in that they're not allowed to give it to more than three people, so I don't know quite what they would have done.'

MONTAGNIER

vs

GALLO

FEUDING PARTIES
Luc Montagnier (born 1932)
– virologist
vs
Robert Gallo (born 1937)
– virologist

DATE
1984–91

CAUSE OF FEUD
Who discovered HIV?

The announcement of the 2008 Nobel Prize for medicine reopened old wounds in the HIV-research community. Half of the prize went to French researchers Luc Montagnier and his colleague Françoise Barré-Sinoussi, for their identification of the virus that causes acquired immune deficiency syndrome (AIDS). Snubbed by the Nobel committee, however, was the American Robert Gallo, the initial claimant to the title 'The Man Who Discovered HIV'. For a period in the 1980s their feud over priority retarded the progress of AIDS research and cost lives, in what science journalist Andy Coghlan calls 'one of the tackiest sagas in the history of medicine'.

The hunt for HIV

In 1981, physicians around the world became alarmed at growing evidence of a new and deadly plague that particularly affected intravenous drug users and gay men. By 1982 the plague had a name – AIDS – and it was apparent that it spread through contaminated blood. But with no idea what was causing the disease, let alone how to test for it, health agencies had no way to check whether their blood supplies were safe, or whether they might be inadvertently infecting unsuspecting patients. Intensive research efforts to identify the cause of AIDS were launched, with the Pasteur Institute in France and the US National Cancer Institute (NCI) in the vanguard.

American virologist Robert Gallo of the NCI had accomplished pioneering work on retroviruses – insidious pathogens that can insert their genetic material into the DNA of a host cell and turn it into a virus-factory – identifying a type of virus he called human T-cell leukaemia/lymphoma virus, or HTLV. His influential voice suggested that a similar virus might be causing AIDS.

Meanwhile, in France, virologist Luc Montagnier had been called in to help isolate viruses from samples of tissue taken from AIDS sufferers. Early in 1983, Montagnier's team isolated a retrovirus from the lymph node of a fashion designer called Frederic Brugiere. They had it photographed under an electron microscope and submitted a paper to *Science* that was published on 20 May. Although he did not name the virus,

TIMELINE

23 April 1984 – US Department of Health, with Gallo in attendance, announces at press conference the discovery of the AIDS virus, applies for US patent for a test

4 May 1984 – Gallo publishes paper in *Science* describing HTLV-III (human T-cell leukaemia/lymphoma virus)

1981: alarm over large numbers of mystery deaths due to immune deficiency

1982: first use of term AIDS (acquired immune deficiency syndrome)

1959: first known case of AIDS

1985 – Pasteur Institute launches legal action over credit and test royalties

1992 – Gallo found guilty of misconduct; charges overturned on appeal

1991 – Gallo admits HTLV-III is same as LAV, blames contamination

1980   1985   1990

1983
20 May – Montagnier's French team publish first paper on unknown virus, later called LAV (lymphadenopathy associated virus) and then HIV (human immunodeficiency virus)
17 July – Montagnier gives Gallo an isolate of LAV
15 September – French team apply for UK patent on test for LAV
23 September – Montagnier gives Gallo second isolate of LAV; Gallo team sign contract promising not to exploit it for commercial purposes
December – French team apply for US patent

1987 – Ronald Reagan and Jacques Chirac broker agreement to settle dispute

2008 – Montagnier wins Nobel Prize; Gallo overlooked

▼ *Township near Cape Town* Poverty and pollution are blamed by 'HIV-deniers' as the real causes of Africa's AIDS epidemic.

Montagnier heeded the advice of the esteemed Gallo and suggested it might be a form of HTLV, although it looked nothing like one. Later, Montagnier would name this virus lymphadenopathy associated virus (LAV), and later still it would be renamed human immunodeficiency virus, or HIV.

Gallo's influence and the respect his opinion commanded among his peers meant that the true and unique nature of LAV was largely ignored; the French paper was overshadowed by one by Gallo in the same issue of *Science*, directing attention to HTLV as the probable cause of AIDS. Talking in 1987, virologist Abraham Karpas claimed this was a tragic mistake: 'A full year was wasted. In that time many lives could have been saved, many infections could have been prevented. Gallo's preoccupation with HTLV as the cause of AIDS led many people in the wrong direction at a critical times in AIDS research.' Similarly, Don Francis of the US Centers for Disease Control (CDC) blamed Gallo for causing 'tremendous confusion'.

Truth and consequences

Unaware that the culprit had already been unearthed, Gallo continued his researches. In July 1983, Montagnier delivered a sample of the LAV isolate to Gallo so that he could test it. The American had little idea how significant the sample was, admitting, 'Was I so excited to get the thing? No, I really wasn't. I put it in my freezer and I went out and played volleyball.' In September of that year, shortly after the French team had applied for a UK patent for a test for LAV (a test that could be worth hundreds of millions of dollars when commercially exploited), Montagnier sent Gallo a further sample, along with a contract stipulating that it not be exploited for commercial purposes. A US patent application by the French followed in December.

Gallo had plenty of samples of his own to test, and in April 1984 word leaked out that he had successfully identified a virus that was common to all the AIDS samples he had looked at. Still adhering to his HTLV hypothesis, Gallo called it HTLV-III. On 23 April, the US Department of Health, with Gallo in attendance, held a press conference trumpeting the new discovery. That same day they applied for a patent for a test for the new virus. In May, Gallo published a paper in *Science* describing HTLV-III, complete with a photo. The French team and other researchers smelled a rat: HTLV-III looked remarkably like LAV.

'It could have happened differently, but everybody has their personality.'

LUC MONTAGNIER, *c.* 1987

A row broke out over priority, while in 1985 the French sued for test royalties. The conflict escalated to the highest level until, in a remarkable development, US president Ronald Reagan and French prime minister Jacques Chirac brokered an out-of-court settlement in which credit was shared between Gallo and Montagnier. In a co-authored 1988 article in *Scientific American*, the men wrote: 'Thus contributions from our laboratories in roughly equal proportions had demonstrated that the cause of AIDS is a new human retrovirus.' The French eventually won a US$6 million settlement in the dispute over test royalties.

But the story did not end there. Controversy over whether Gallo's HTLV-III was the same as LAV rumbled on until, in 1991, Gallo came clean in a letter to *Nature*, admitting that the two were the same; it later transpired that they had even come from the same patient, Brugiere. The mistake 'appeared' to have arisen from contamination of Gallo's cell cultures by the French virus. In 1992, a review panel of the National Academy of Sciences accused him of 'intellectual recklessness of a high degree', and he was found guilty of scientific misconduct by the Office of Research Integrity of the US Public Health Service. All charges were overturned on appeal, but there is little doubt the controversy besmirched Gallo's name to the point of excluding him from Nobel considerations, and had far worse consequences besides. According to science journalist Steve Connor, one of the first to investigate the scandal: 'blood went unscreened, spreading the virus further, a climate of distrust among the key scientists developed, spilling over into the courts. Scientific collaboration itself joined the list of victims.'

The controversy also threatens to overshadow the immense contributions Gallo has made to AIDS research. He made several of the breakthroughs that allowed definitive identification of LAV as the agent of AIDS, and was responsible for many subsequent advances. He and Montagnier are now close colleagues and friends, the Frenchman lamenting the decision of the Nobel committee to exclude Gallo:

'it is certain that he deserved this as much as us two.' Gallo himself was gracious about the snub, admitting that it was 'a disappointment', but praising the winners: 'I am pleased my long-time friend and colleague Dr Luc Montagnier, as well as his colleague Francoise Barre-Sinoussi, have received this honor. I was gratified to read Dr Montagnier's kind statement this morning expressing that I was equally deserving. I am pleased that the Nobel Committee chose to recognize the importance of AIDS with these awards.'

AFRICAN AIDS CONTROVERSIES

Africa has borne the brunt of the AIDS epidemic, and it has also become the battleground for some vicious and deadly feuds over the causes and treatment of the disease. A cadre of HIV/AIDS denialists, based in South Africa where they received support at the highest level from former president Thabo Mbeki, have claimed that AIDS is not a disease caused by HIV, but a consequence of poverty and malnutrition, a reflection of social injustice and borderline racism in the developed world. They oppose use of the highly effective cocktail of anti-retroviral drugs that, while it does not cure AIDS and can have nasty side effects, is proven to prolong lives and reduce HIV transmission rates, and advocate 'natural' treatments, such as vitamin supplements. *(continued overleaf)*

▶ **Quite contrary.**
Controversial biology professor Peter Duesberg in his lab at Berkeley University in 1992. Duesberg won acclaim for his work on the genetics of cancer before alienating mainstream opinion with his views on HIV/AIDS.

One of the most prominent HIV/AIDS denialists is Peter Duesberg, a formerly highly respected and much garlanded professor of molecular biology, who claims the AIDS epidemic is a myth, a scam perpetrated by scientists and the pharmaceutical industry to build careers, win funding and rake in huge profits. Duesberg gained publicity from a series of articles in the British newspaper *The Sunday Times*, articles blasted by the journal *Nature* as 'seriously mistaken, and probably disastrous'. Harvard professor Max Essex, a former friend of Duesberg's, claims that history will judge him as either 'a nut who is just a tease to the scientific community' or an 'enabler to mass murder'. Among Duesberg's disciples was the high-profile HIV/AIDS denialist and activist Christine Maggiore, who refused to take ARV medication or other anti-HIV precautions despite being pregnant and campaigned to convince other HIV-positive mothers to do the same. One of her daughters subsequently died of AIDS-related illness, aged three years, and in 2008 Maggiore herself died of pneumonia (a common cause of death in AIDS sufferers).

In 2000, Mbeki convened a Presidential Advisory Panel on HIV and AIDS, which included Duesberg, and later that year he publicly rejected the belief that AIDS is caused by HIV. As a result, provision of drug therapy to AIDS sufferers was interrupted, even after drug companies offered free medicine, and education programmes to limit the spread of HIV were hampered. According to a 2008 study by Pride Chigwedere and colleagues from the Harvard School of Public Health in Boston, the policies of Mbeki's government probably led to 330,000 unnecessary deaths and the infection of 35,000 more babies with HIV than necessary. The researchers say their estimates were based on 'very conservative assumptions', and conclude: 'In the case of South Africa, many lives were lost because of a failure to accept the use of available ARVs to prevent and treat HIV/AIDS in a timely manner.'

Mattias Rath is another high-profile AIDS denialist. He sells nutritional supplements around the globe through the Dr Rath Foundation, and claims that while conventional medicines are toxic and disease-causing, his supplements can treat a wide range of severe diseases, including cancer, heart disease, stroke and, perhaps most controversially, AIDS. In South Africa, Rath had considerable success promoting his supplements as an alternative to ARV treatment for AIDS, propounding a message that UK HIV/AIDS expert Professor Brian Gazzard describes as 'extremely harmful'. According to Mark Wainberg, director of the McGill AIDS Centre in Montreal, 'It is clear that he [Rath] has done enormous harm to people with HIV.' In 2008, South African courts found that a

▲ **Activists** *from South Africa's Treatment Action Campaign for better treatment for HIV sufferers, demonstrating during a court case involving Matthias Rath.*

trial of Rath's nutritional supplement VitaCell in treatment of AIDS was illegal; some of the participants in the trial died and relatives of some of them said that they had been told to stop using conventional medicines.

Further controversy flaired when Rath's colleague Anthony Brink, a former spokesman for the Dr Rath Foundation, attempted to have South African AIDS activist Zackie Achmat indicted for genocide, filing a complaint that called for him to be permanently confined 'in a small white and concrete cage, bright fluorescent light on all the time to keep an eye on him', in which he should be force-fed AIDS medication, which, 'if he bites, kicks and screams too much, [should be] dripped into his arm after he's been restrained on a gurney with cable tied around his ankles, wrists and neck.' When science journalist Ben Goldacre of the *Guardian* newspaper and 'Bad Science' blog drew attention to some of these controversies, Rath sued the newspaper for libel. In September 2008 he dropped the case and was ordered by a UK court to pay £220,000 in initial costs.

SCIENCE AND POLITICS

Science and politics have been intertwined since before the start of the Scientific Revolution in the 17th century. The Royal Society of London lobbied hard to win royal patronage, knowing that it would secure their status and, crucially, boost their funding. Similar situations arose in France, where the Académie des Sciences was even more directly connected with the monarch than its English counterpart, and in the Holy Roman Empire, where the favour of emperors such as Rudolf II had briefly made Prague the intellectual epicentre of the Early Modern World. Natural philosophers were in the vanguard of radical and revolutionary movements, such as the Rosicrucians and later the American and French revolutions.

As scientific institutions grew and science became grander, more ambitious and more expensive, so the state took a greater role and politics became more important in determining where funds were allocated and in which direction research should proceed. These trends were intensified during periods of war, as national scientific capabilities were directed towards improving the ability to wage war and, particularly, to the development of weapons technology. The First World War, for instance, saw great advances in chemistry and radio, while the Second World War drove advances in radar, sonar, electronics, computers and atomic energy. The 20th century also saw science subordinated to ideology, as both the Nazis and the Soviets looked to science to justify their world views, while also attempting to reshape science in their own image. The Nazis, for instance, employed spurious anthropology and genetics to bolster their views on race, while the Soviets suppressed some fields of science, such as Mendelian genetics, while elevating other areas, such as Lysenkoism. Lysenkoism, based on the ideas of pseudoscientific charlatan Trofim Lysenko (1898–1976), had enduring and terrible consequences for both Soviet agriculture, leading to crop failures and famines in which many starved, and Soviet science —

Senator Arlen Spector, watched by actor and Parkinsons disease research activist Michael J. Fox and other politicians, campaigning for legalization of stem cell research – a 'hot button' issue in the culture wars of America.

not only was genetics declared a 'bourgeois pseudoscience' under Stalin, many scientists who opposed Lysenko, or who simply studied in the wrong field, were sent to labour camps.

Attempts to impose political ideology on science continue today. For instance, recent years saw the Bush administration in America pursuing the agenda of its core conservative supporters in the way it treated science. Some areas of research — most notoriously research on stem cells — were stripped of federal funding or blocked altogether. More generally, the Republicans, both in the Bush administration and, more emphatically, in the last presidential campaign, adopted an anti-intellectual stance and tried to portray science and scientists as liberal and irrelevant. For instance, a research project into grizzly bear genetics in Montana came under sustained attack, with presidential candidate John McCain describing federal spending of $3 million on the project as 'criminal' and 'unbelievable' — although scientists protested that the study had been useful and quite important. Climate science has been a particular battleground, with state-sponsored science in America strongly affected by political concerns — many climate scientists claimed that their reports were rewritten for political reasons.

Greenpeace activists protest outside the German Reichstag in 2004, opposing the granting of patents on human genetic material. Ethical issues in science, particularly applied science, are legitimately political questions: it is the job of science to discover what is possible, but of politicians and judges to decide what is permissible.

Science itself likes to claim that it is apolitical; an objective process in which personal or institutional prejudices, opinions, etc. play no part. But since the 1960s and '70s this claim has been challenged by philosophers and historians, such as Thomas Kuhn and Pierre Durhem, and investigated in detail by sociologists and anthropologists, such as Peter Galison, who explored the social world of 'big science' (projects such as CERN — the world's largest particle physics laboratory — or the Human Genome Project [*see* pages 156–161]). A post-modern, post-positivist view emerged that science was as political as any other field of human endeavour.

VENTER
vs
THE HUMAN GENOME PROJECT

FEUDING PARTIES
J. Craig Venter (born 1946)
– geneticist
vs
Francis Collins (born 1950)
– geneticist;
John Sulston (born 1942)
– molecular biologist

DATE
1991–*c.*2001

CAUSE OF FEUD
Sequencing of the human
genome

The sequencing of the entire human genetic code is one of the greatest collective endeavours in scientific if not human history. Yet what was supposed to be a model of international collaboration became what *Newsweek* called 'the biggest scientific grudge match since the space race'.

Skimming the cream

Trouble started in 1991 when Craig Venter, a maverick scientist with a distrust of government interference, riled senior colleagues by publicly suggesting that the Human Genome Project (HGP) should focus exclusively on those parts of the human genomes that coded for genes, and ignore the vast stretches of what is known as 'junk DNA'. Venter, then working for the National Institutes of Health (NIH), also endorsed patenting of genes as they were discovered, and some of his bosses initially agreed. The project leader at the time, James Watson, was furious; he dismissed Venter's 'cream-skimming approach', saying it could be done by 'virtually any monkey', and denounced the rush to patent

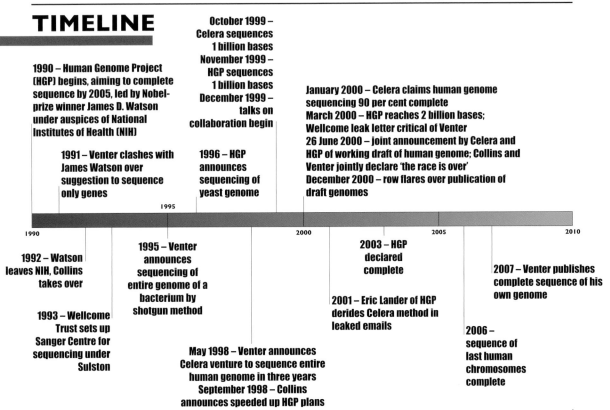

TIMELINE

October 1999 – Celera sequences 1 billion bases
November 1999 – HGP sequences 1 billion bases
December 1999 – talks on collaboration begin

1990 – Human Genome Project (HGP) begins, aiming to complete sequence by 2005, led by Nobel-prize winner James D. Watson under auspices of National Institutes of Health (NIH)

1991 – Venter clashes with James Watson over suggestion to sequence only genes

1996 – HGP announces sequencing of yeast genome

January 2000 – Celera claims human genome sequencing 90 per cent complete
March 2000 – HGP reaches 2 billion bases; Wellcome leak letter critical of Venter
26 June 2000 – joint announcement by Celera and HGP of working draft of human genome; Collins and Venter jointly declare 'the race is over'
December 2000 – row flares over publication of draft genomes

1995

1990   2000   2005   2010

1992 – Watson leaves NIH, Collins takes over

1995 – Venter announces sequencing of entire genome of a bacterium by shotgun method

2003 – HGP declared complete

2007 – Venter publishes complete sequence of his own genome

1993 – Wellcome Trust sets up Sanger Centre for sequencing under Sulston

2001 – Eric Lander of HGP derides Celera method in leaked emails

2006 – sequence of last human chromosomes complete

May 1998 – Venter announces Celera venture to sequence entire human genome in three years
September 1998 – Collins announces speeded up HGP plans

THE HUMAN GENOME PROJECT

The genetic code of a human being is 3 billion letters long; in the 1980s it was the final frontier of biology. The project to sequence almost every letter of this code was initially regarded as a wildly ambitious act of scientific folly. Critics pointed to the fact that science had yet to sequence successfully a humble bacterium, let alone a higher vertebrate. The technology to sequence short lengths of DNA existed, but the process was laborious, slow and very expensive. The time, manpower and money that would be necessary would divert resources from other areas of science, and what little was known about the human genome suggested that most of the effort would be wasted on sequencing vast lengths of 'junk DNA', which does not contain genes and is of unknown function.

In 1985, University of California chancellor Robert Sinsheimer convened a gathering of top geneticists to consider launching an assault on the ultimate prize in genome sequencing. They concluded the project was not feasible, but the genie could not be put back in the bottle. Charles DeLisi of the US Department of Energy (DoE) decided it could be just the grand project the DoE needed now that demand for its atomic-bomb-making skills was in decline.

In 1986, DeLisi instituted another workshop, but there was still widespread scepticism, even ridicule, of the ambitious scheme. One early critic, geneticist David Botstein, described it as 'a scheme for unemployed bombmakers'. More seriously, scientists were concerned about the resources the project would gobble up, and how this would impact on other sciences, particularly biological ones.

Another panel, convened by the National Research Council in 1988, mapped out a way forward that mollified the sceptics, with a phased approach that tackled the genomes of simpler organisms first. Work on the human genome would begin by first mapping the chromosomes, which could highlight disease genes of immediate biomedical importance. Detailed sequencing could come later as technology advanced. The human genome project (HGP) was now go.

By 1990, the US National Institutes of Health had wrested control of the project. As top overseas institutions came on board, notably Britain's Wellcome Foundation Sanger Research Laboratory under John Sulston, the HGP seem destined for a calm progression to its forecasted 2005 completion date. Few suspected the storms that lay ahead.

> ## *'If there is anything worth doing twice, it's the human genome.'*
>
> DAVID HAUSSLER, PROFESSOR OF BIOMOLECULAR ENGINEERING, JULY 2000

genes as 'sheer lunacy', telling Congress, 'I am horrified'. Watson lost his job at NIH as a result, while Venter had already jumped ship to set up a new venture with private money.

Venter made an even bigger splash in 1998 when he set up Celera Genomics, a privately funded company, and announced to the world that he was going to use masses of hi-tech sequencers and supercomputers to sequence the human genome, using a controversial technique known as whole genome shotgunning. Although he would not try to patent any discoveries, Venter did plan to profit from them. Worst of all, he claimed he would accomplish all this faster and cheaper than the slow and steady HGP.

Venter's bombshell sparked a 'genome war' and made him public enemy number one for many publicly funded scientists. In the course of an increasingly bitter feud with his rivals he has been called 'Hitler', 'a self-aggrandizing pain in the arse' and 'an opportunistic maniac'. Initial reactions ranged from biologist Maynard Olson's derision of 'science by press release' to more accusations of 'cream-skimming' – 'It's clearly an attempt to short-circuit the hard problems and defer them to the [research] community at a very substantial cost,' Robert Waterstone told the journal *Science*. There were also serious anxieties about the dangers of commercializing the human genome: 'If global capitalism gets complete control of the human genome,' warned John Sulston, 'that is very bad news indeed. I do not believe it should be under the control of one person.' Venter dismissed the criticism: 'I have been a thorn in their side for some time because I keep coming up with breakthroughs. Having a rival in any sense is unacceptable to them.'

Pizza treaty

With Celera, Venter had turned the sequencing of the human genome into a race, and it quickly escalated to an all-out war. Venter did his best to maintain his cool in the face of continued sniping, admitting 'If I were on the other side of this, I would feel upset and threatened, too.' But at times he was exasperated: 'Why should I play by their rules

▲ **Brokering the peace.** *US President Bill Clinton, flanked by J. Craig Venter (L) and Francis Collins (R), announces the completion of the first draft of the human genome in 2000.*

when I am not getting a cent of federal money?' he complained in a 1999 interview. He was also capable of a fair degree of egotism: 'Is my science of the level consistent with other people who have gotten the Nobel? Yes.'

The unseemly row stoked the ire of then president Bill Clinton, who told aides to get the warring scientists in line. Peace talks began in late 1999 but were almost derailed in March 2000 when Wellcome leaked a letter from Sulston to Venter citing irreconcilable differences. Venter described the incident as 'a lowlife thing to do'. Efforts to reconcile the parties continued, however, and over beer and pizza in a basement, a deal was thrashed out. Bill Clinton and the British Prime Minister at the time, Tony Blair, would host an announcement at the White House in June, at which Venter and Collins would announce that the HGP and Celera had jointly achieved a first draft of the human genome.

Many scientists were relieved. 'Everyone is sick to death of it,' said Richard Gibbs, of the Baylor College of Medicine in Houston, 'If this ends the horse race, science wins.' Collins voiced similar sentiments: 'Ten, fifteen years from now, nobody is going to care about all this fuss and bother … all this back and forthing over who did what and what strategy was used and which money was public and which was private is probably going to sink below the radar screen. And hallelujah.'

But the fighting was not over yet. A new row broke out over how and where the results would be published, and with much of the donkey work still to do there was plenty of time for recriminations. In May 2001, emails surfaced from Eric Lander of the Whitehead Institute, a leading member of the HGP, disparaging Celera's 'shotgun' method as 'a flop. No ifs, ands or buts. Celera did not independently produce a sequence of the genome at all. It rode piggyback [on the HGP].' Venter was not amused: 'We think there is zero legitimacy to anything Eric is saying, and we don't understand why he is saying it.' By 2003, the project was largely complete, and the main protagonists have mostly moved on to other things (Venter, for instance, is designing new life forms to solve the energy crisis).

▼ **Reading the code.** *On this printout from a sequencing machine, the four different colours correspond to the four types of nucleotide that make up the letters of the DNA code.*

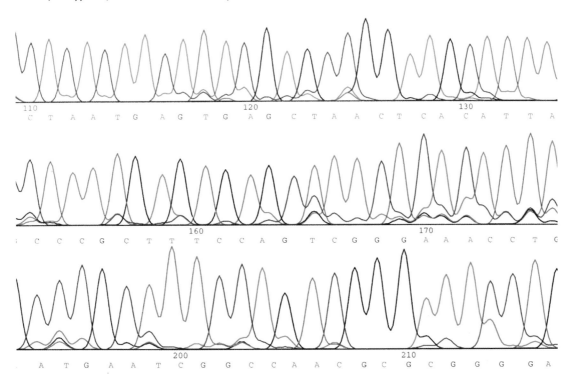

PART FOUR
PHYSICS, ASTRONOMY and MATHS

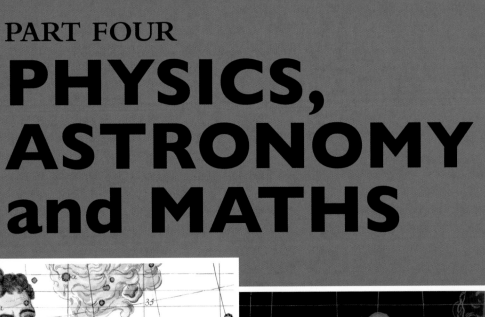

TYCHO
vs
URSUS

FEUDING PARTIES
Tycho Brahe (1546–1601)
– Danish nobleman, Imperial
Astronomer
vs
Nicolaus Reimarus 'Ursus'
(1551–1600) – German
swineherd, servant, Imperial
Astronomer

DATE
1584–1600

CAUSE OF FEUD
The make-up of the solar system

The last few years of Tycho Brahe's life were consumed with an extraordinarily vicious spat with a rival astronomer he accused of stealing his intellectual property and publishing it as his own. The alleged culprit, Nicolaus Reimarus, known as the 'Bear', or *Ursus* in Latin, aroused particular spleen because of his lowborn origins: supposedly he had started life as a lowly swineherd, before graduating to the service of noblemen, including that of Tycho's friend Eric Lange. In 1588, shortly after Tycho had published his 'system of the world' (a model of the universe), Ursus published one very similar. Ursus' book *Fundamentals of Astronomy* helped secure for him the post of Imperial Astronomer to Emperor Rudolf II in Prague, but also unleashed a wave of fury from Tycho.

'Long-nosed Ursus'

In a series of letters, Tycho later set out the case against his despised opponent. Lange, accompanied by the 'rotten and sycophantic' Ursus, had visited Tycho at his observatory, Uraninborg, on the island of Hven, in September 1584. After dinner, Tycho outlined for Lange his criticisms of both the prevailing Ptolemaic system, which was geocentric (with the Earth at the centre of the Solar System and the universe), and the newer Copernican system, which was heliocentric (with the Sun at the centre). His unequalled catalogue of astronomical observations had convinced him they were both wrong, and he sketched out his own geoheliocentric system (with the Sun and Moon orbiting the Earth and the other planets orbiting the Sun), using chalk on a tablecloth. Suspicious of intellectual theft, Tycho erased the diagram. 'But afterward that long-nosed Ursus,' he recounted, 'sniffing that we had hidden this from him, conceived some idea of my hypotheses either from the traces left on the tablecloth, or from a certain paper that used to be in my study.'

In practice, there were some differences between the systems of the two men – in Ursus' version the Earth had a daily rotation and the orbit of Mars did not cross that of the Sun – but for Tycho this was the smoking gun in the case. Writing to the Imperial Physician, Hagecius, in 1592, he enclosed 'the very piece of paper … from which he doubtlessly took his idea when he was here'. It was an 'erroneous diagram … which had been drawn in this way through some inadvertence, and, for that reason, was thrown away somewhere among my papers as inaccurate … by imitating this sketch, these orb[it]s are incorrectly drawn in his diagram'.

In 1597, Ursus published a counterattack described by the *Journal for the History of Astronomy* as 'savage … [Ursus] was shrewd in his choice of targets: Tycho's duel-disfigured nose; the propriety of his marriage; [his] vanity.' In 1599, Tycho launched legal action against Ursus, who fell ill and died in August 1600. Tycho was implacable, however. He succeeded in having all stocks of Ursus' defamatory book burned and pursued official condemnation of his enemy until his death in October 1601.

▶ **An astronomical arc** One of the instruments Tycho used at his observatory at Uraninborg, where he compiled the greatest astronomical catalogue yet created.

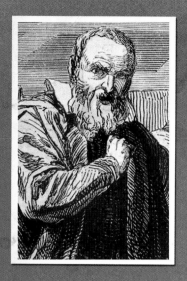

GALILEO

vs

POPE URBAN

FEUDING PARTIES
Galileo Galilei (1564–1642)
– natural philosopher and
astronomer
vs
Urban VIII, aka Maffeo Barberini
(1568–1644) – Pope

DATE
1632–42

CAUSE OF FEUD
Whether the Earth moves

In its treatment of Galileo and its apparent pig-headed refusal to look through a telescope, the Catholic Church scored a lasting own goal. Ever since, the clash between the Florentine astronomer and the Pope has become the ultimate exemplar of dogmatism suppressing reason. The story begins in 1609, when Galileo built a telescope and used it to make a number of startling discoveries – particularly the moons of Jupiter, which orbited the planet like a mini solar system – that undermined the traditional Ptolemaic, geocentric system and lent support to the Copernican, heliocentric system. Previously, Galileo had been circumspect in his enthusiasm for Copernicus, writing to the German astronomer Johannes Kepler in 1597: 'I have not dared until now to bring my reasons and refutations into the open, being warned by the fortunes of Copernicus himself, our master, who procured for himself immortal fame among a few but stepped down among the great crowd.'

'The pertinacity of the asp'

Now, it seemed the truth was plain to see for anyone who looked through a telescope, and he was determined to spread the word, only to be surprised at the resistance he encountered. The evidence of the senses along with passages in the Bible clearly

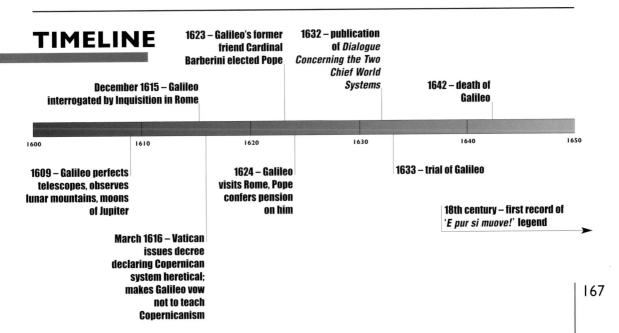

TIMELINE

1623 – Galileo's former friend Cardinal Barberini elected Pope

1632 – publication of *Dialogue Concerning the Two Chief World Systems*

December 1615 – Galileo interrogated by Inquisition in Rome

1642 – death of Galileo

1600   1610   1620   1630   1640   1650

1609 – Galileo perfects telescopes, observes lunar mountains, moons of Jupiter

1624 – Galileo visits Rome, Pope confers pension on him

1633 – trial of Galileo

18th century – first record of 'E pur si muove!' legend

March 1616 – Vatican issues decree declaring Copernican system heretical; makes Galileo vow not to teach Copernicanism

167

> ‘I curse the time devoted to these studies in which I strove and hoped to move away somewhat from the beaten path. I repent having given the world a portion of my writings; I feel inclined to consign what is left to the flames and thus placate at last the inextinguishable hatred of my enemies.’

GALILEO, 1633

seemed to favour the doctrine of terrestrial immobility, and Galileo was stumped, writing to Kepler in 1610: 'My dear Kepler, what would you say of the learned here, who, replete with the pertinacity of the asp, have steadfastly refused to cast a glance through the telescope? What shall we make of this? Shall we laugh, or shall we cry?'

Galileo circulated a letter in which he argued that Scripture is not intended to be taken literally, and cannot be looked to for the answers to everything: 'Who would dare assert that we know all there is to be known?' This only served to stoke the ire of reactionary clergymen, including Dominican friar Father Lorini who, in February 1615, wrote to the Inquisition to complain: 'the letter contains many propositions which appear to be suspicious or presumptuous, as when it asserts that the language of Holy Scripture does not mean what it seems to mean … that the Holy Scriptures should not be mixed up with anything except matters of religion.'

◀ **Statue of Galileo** in Florence, Italy, the city where he grew up, and where he would ultimately live out the last years of his life, under virtual house arrest.

GALILEO GALILEI

HAS THE CHURCH BEEN HARD DONE

The Catholic Church has paid for its condemnation of Galileo, and misconceptions about the trial abound. The astronomer was not, for instance, tortured or burned at the stake as many believe, and the story that he muttered '*E pur si muove!*' ('And yet it moves'), even as he got back to his feet after kneeling to make his abjuration of terrestrial motion, is a legend that first surfaced in the 18th century. Apologists for the Church, such as J. Gerard of the *Catholic Encyclopaedia*, contend that 'it is obviously absurd to maintain the idea that the opposition of the ecclesiastical authorities was grounded, as is constantly assumed, upon a fear lest men should be enlightened by the diffusion of scientific truth.' Gerard argues that at the time there were legitimate doubts about the Copernican system, which had not been definitively proved by Galileo, and that the astronomer spent only a few weeks in the custody of the Inquisition, during which time he was well treated. The Church has worked to rehabilitate Galileo (and itself), and in 2008 the Vatican unveiled plans to erect a statue of the great man: 'to close the Galileo affair and reach a definitive understanding not only of his great legacy but also of the relationship between science and faith'.

▶ *Galileo's telescopes.* Galileo designed and built his own telescopes and with them he was able to use the evidence of his own eyes to challenge ancient and Biblical authority.

Galileo was now in hot water. That April, the powerful Cardinal Bellarmine opined: 'to affirm that the Sun, in its very truth, is at the centre of the universe … is a very dangerous attitude … calculated … to injure our faith by contradicting the Scriptures.' Galileo pleaded that his evidence not be dismissed 'without understanding it, without hearing it, without even having seen it'. The Bible was intended to teach men to go to heaven, he asserted, not to teach them how the heavens go.

It was to no avail. On 23 February 1616, Galileo's heliocentric propositions were unanimously declared 'foolish and absurd' and 'formally heretical'. He was admonished 'to abandon the said opinion … that the Sun is the centre of the world and immovable and that the Earth moves'. From now on, Galileo could only teach Copernicanism as a 'hypothesis', not as fact. But there was confusion over whether he had been formally 'commanded and enjoined' – he said no, the Inquisition said yes. The legal niceties would later become important.

Ventured to meddle

His run-in with the Inquisition silenced Galileo for a number of years, but in 1623, after his friend Cardinal Barberini was elected Pope Urban VIII, he received a letter from the papal private secretary: 'If you would resolve to commit to print those ideas that you still have in mind, I am quite certain that they would be most acceptable to His Holiness … You should not deprive the world of your productions.'

Encouraged, Galileo embarked on his *Dialogue Concerning the Two Chief World Systems*, in which the Ptolemaic, geocentric corner is fought by a simple-minded stick-in-the-mud named Simplico. The book was published in 1632, but the Pope did not react as expected, issuing dark warnings to an intermediary: 'Your Galileo has ventured to meddle in things that he ought not

◀ **By the moons of Jupiter.** *A page from Galileo's notebook recording his observations of Jupiter and its satellites (now known as Galilean moons) in 1610.*

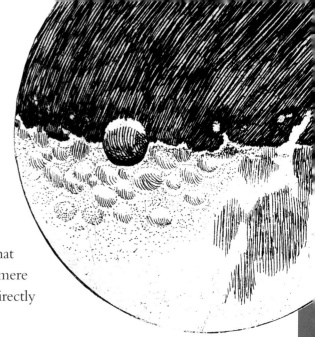

▶ **Moonscape.** *Galileo's 1611 drawing showing the Moon's surface in unprecedented detail. His discoveries proved that there were more things in heaven than dreamt of by the ancients.*

and with the most grave and dangerous subjects that can be stirred up these days.' The Pope felt that Galileo had betrayed him by not sticking to the 'mere hypothesis' line, and was even lampooning him directly with the character of Simplico.

Urban called in the Inquisition, which promptly declared that Galileo had violated the terms of his 1616 injunction – an injunction that he claimed was never issued to him. He was summoned to Rome to appear before the Inquisition, arriving in February 1633, ill and broken. It was reported that 'for two nights continuously [he] cried and moaned in sciatic pain; and his advancing age and sorrow.'

Found guilty of presenting the Copernican model as 'a physical reality', he was urged to admit error in return for a lenient sentence. He did so, a decision that has sometimes led to criticism that he was 'selling out', rather than sticking to his beliefs whatever the cost. If he had any doubts about what that cost might be, however, he need look back only 32 years to the fate of Giordano Bruno, who was tried for a similar crime to Galileo's – asserting that the Earth was not the static centre of the universe – in the very same rooms in Rome and by the very same Cardinal Bellarmine as Galileo. For his refusal to abjure his beliefs, Bruno was burned to death at the stake.

Despite his admission, Galileo was found guilty in June of being 'vehemently suspected of heresy', 'condemned to imprisonment at the pleasure of the Holy Congregation [the Inquisition]' and to recite the Seven Penitential Psalms once a week for three years. The Dialogues were 'to be prohibited'.

Galileo was eventually allowed to move to a small farmhouse of his own, where he lived until his death in 1642, but the damage had been done. The Church's insistence on privileging dogma over evidence and argument was responsible, according to some, for ending the Renaissance in Italy and surrendering forever the country's leading status in world science.

SCIENCE AND THE CHURCH

Debates about science and religion are mainly framed with reference to Christianity and particularly the biggest Christian institution, the Catholic Church. This is partly because modern science developed largely in the West, so that relationships between science and Eastern belief systems are often tangled up with the discourse of colonialism. In the West, there is a readily available narrative, familiar to most, of a centuries' long war between faith and reason, religion and science, with the names Copernicus, Galileo (*see* pages 166–171) and Darwin (*see* pages 14–21 and 38–47) signifying important battles. In this narrative, the forces of reason and truth have slowly but inevitably prevailed over the dark forces of ignorance and dogma. Not surprisingly, many religious commentators object to this narrative and argue that it is false. So, where does this narrative come from?

Steeped in blood

The simple answer is the late 19th century, which saw an outbreak of reductionist, anti-religious histories, most notably John William Draper's 1874 *History of the Conflict between Religion and Science*, which was soon trumped by Andrew Dickson White's *A History of the Warfare of Science and Theology in Christendom* in 1896. These books recounted the now familiar narrative of Galileo, et al., painting religion, specifically the Vatican, as the bad guys in an epic conflict between darkness and light. 'As to Science,' wrote Draper, 'she has never sought to ally herself to civil power. She has never attempted to throw odium or inflict social ruin on any human being. She has never subjected any one to mental torment, physical torture, least of all to death, for the purpose of upholding or promoting her ideas. She presents herself unstained by cruelties and crimes. But in the Vatican— we have only to recall the Inquisition—the hands that are now raised in appeals to the Most Merciful are crimsoned. They have been steeped in blood!'

This tradition was continued by T.H. Huxley (*see* pages 38–47), who helped to form the now dominant narrative of the Church's reaction to evolutionary theory. These men often had their own motivations for promulgating such a polarized view. Huxley, for instance, was a convinced agnostic and wanted to challenge the authority of the Church in social as well as intellectual areas.

The Book of Nature

Modern scholars of the relationship between science and religion tend to present a much more nuanced view, based in part on the heterogeneity of both scientific and religious attitudes. To talk about a single Church with a single reaction to a monolithic Science is false. In the specific example of Copernicanism, for instance, reactions were extremely mixed and developed over a long period — Copernicus himself was a Catholic priest, and religious reaction to his heliocentric theory ranged from enthusiastic to curious to hostile.

In fact, there is a strong argument that the Church was instrumental in giving birth to scientific enquiry. The roots of Western philosophy were in the Church; Augustine, for instance, was clear that scripture was not the only possible source of information, and there was a strong tradition in Christianity that the Book of Nature was to be read alongside the Book of Scripture. Many of the great figures in Early Modern science saw it as their sacred duty to read the Book of Nature, and doing this through scientific investigation was a kind of

In defence of authority. The Reformation was a challenge to the orthodoxy and thus the authority of the Church. Accordingly, the Counter-Reformation, spearheaded by the Inquisition, shown here burning heretics in Spain, was keen to reassert authority by opposing heterodoxy — including new natural philosophies.

sacrament. The astronomers Johannes Kepler and Isaac Newton, for instance, were both extremely devout, believing that God revealed his power and wonder through the rational and harmonious (and ultimately mathematical) design of the universe.

Problems arose over the literal truth of some Biblical verses, especially ones concerned with the motion of the Earth and Sun. Joshua 10:13, for instance, says 'So the sun stood still in the midst of heaven', but Isaiah 40:22 speaks of 'the heavens stretched out as a curtain' above 'the circle of the earth'. Galileo ran into trouble when he tried to argue 'There are in Scripture words which, taken in the strict literal meaning, look as if they differed from the truth,' especially when an unscrupulous opponent told the Vatican he had actually written 'There are in Scripture words which are false in their literal meaning.'

A statue of the heretic philosopher Giordano Bruno, erected in Rome in 1889 as a comment on the attitude of the Papacy to freedom of thought by Freeemasons angry at attacks on them by the Church. Bruno was a Copernican who told the Inquisition, 'the theories on the movement of the Earth and on the immobility of the firmament or sky are by me produced on a reasoned and sure basis, which doesn't undermine the authority of the Holy Scriptures.' Bruno was burned at the stake in 1600.

More generally, science and the Church came into conflict over issues such as certainty and doubt. Scepticism was necessary for science, but it constituted a challenge to authority at a time when the Catholic Church had suffered the attacks of the Reformation and felt it had to stand its ground. Even then, the Vatican could be open to science, especially when it had practical purposes. The Vatican Observatory, for instance, traces its history back to 1582, when Pope Gregory XIII convened a committee to advise on calendar reform. According to Pope Leo XIII: 'Pope Gregory XIII ordered a tower to be erected ... and to be fitted out with the greatest and best instruments of the time. There he held the meetings of the learned men to whom the reform of the calendar had been entrusted ... When touched by the rays of the sun that are allowed to enter from above, the designs demonstrate the error of the old reckoning.' In the 18th and 19th centuries, the Vatican founded three observatories, culminating in Pope Leo XIII officially refounding the Vatican Observatory in 1891.

Modern tensions

Protestantism had been generally more conducive to free inquiry, but there was no simple correspondence. Historians of Anglicanism, for instance, contend that Anglican opposition to Darwinism has been grossly overstated ever since Huxley, and certainly today the Anglican position is squarely in favour of science and against recent fundamentalist movements, such as Creationism and Intelligent Design (*see* page 45). Such movements signal a growing tendency in more extreme religious circles (both in Christianity and in Islam) to view science and, particularly, evolutionary science as an atheistic and immoral enemy. Even in the Catholic Church, after a long period of rapprochement during which, for instance, a 1992 Pontifical Commission acknowledged that Galileo was right and blamed the affair on 'tragic reciprocal incomprehension', new tensions are emerging. Whereas Pope John Paul II seemed to have no problems with evolution, there are suspicions that Pope Benedict is pro-Intelligent Design, and that the director of the Vatican Observatory, Father Coyne, lost his job because of his hostility to that pseudo-scientific movement.

Sir John Templeton (1912–2008). A billionaire with a lifelong interest in science and spirituality, his Templeton Prize, the largest annual prize in the world, is awarded to the individual who 'has made an exceptional contribution to affirming life's spiritual dimension, whether through insight, discovery, or practical works.'

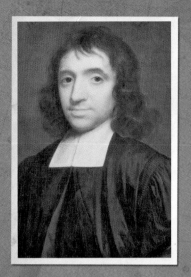

NEWTON

vs

FLAMSTEED

FEUDING PARTIES
Isaac Newton (1642–1727)
– scientist, President of Royal
Society
vs
John Flamsteed (1646–1719)
– Astronomer Royal

DATE
1680–1727

CAUSE OF FEUD
The star catalogue

It was consideration of the Moon (not the famous apple of legend) that first led Newton to formulate his famous law of gravity, which he laid out in his magnum opus, the *Philosophiae Naturalis Principia Mathematica* – 'Mathematical Principles of Natural Philosophy' – usually known as the *Principia*. Yet the Moon was also a source of frustration for Newton. He had demonstrated the theoretical power of his new system of the world by using it to account precisely for the perturbations in the orbit of Saturn and the motion of the tides, but he realized that the greatest demonstration possible would be a complete account of the orbit of the Moon. Unfortunately, the observations of the lunar orbit then available to him were insufficient for his purposes, and his plan to make a 'theory of the Moon' the centrepiece of a second edition of the *Principia* depended on getting hold of raw data from the only man in the world who could supply it: John Flamsteed, the Astronomer Royal.

'For their advantage & your reputation'

Newton and Flamsteed had corresponded years earlier, when two comets seen in rapid succession in the winter of 1680 had provoked each to speculate on the possibility that a single comet, approaching and then receding from the Sun, might be responsible for the sightings. The discussion had helped to develop Newton's thinking on gravity. In 1694, Newton renewed contact, beginning with honeyed words: 'for your observations to come abroad thus with a theory which you ushered into the world ... would be much more for their advantage & your reputation than to keep them private ... For such a theory will be a demonstration of their exactness and make you readily acknowledged the most exact observer that has hitherto appeared in the world.'

Newton could make himself agreeable when he wanted something but he did not suffer fools gladly and was not the most diplomatic of correspondents. Above all, he was demanding, pressing Flamsteed for a flood of observations. The Astronomer Royal,

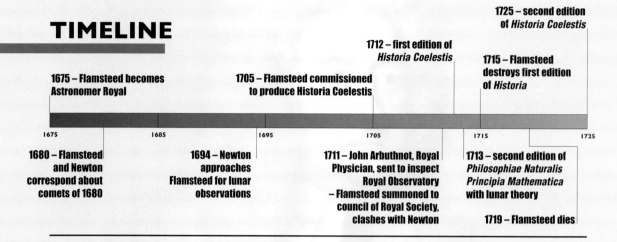

1725 – second edition
of *Historia Coelestis*

1712 – first edition of
Historia Coelestis

1715 – Flamsteed
destroys first edition
of *Historia*

1675 – Flamsteed becomes
Astronomer Royal

1705 – Flamsteed commissioned
to produce Historia Coelestis

1675   1685   1695   1705   1715   1725

1680 – Flamsteed
and Newton
correspond about
comets of 1680

1694 – Newton
approaches
Flamsteed for lunar
observations

1711 – John Arbuthnot, Royal
Physician, sent to inspect
Royal Observatory
– Flamsteed summoned to
council of Royal Society,
clashes with Newton

1713 – second edition of
Philosophiae Naturalis
Principia Mathematica
with lunar theory

1719 – Flamsteed dies

despite his grand title, was poorly remunerated and struggled to make ends meet, and he relied on assistants to help compile his data. Trying to be helpful, Flamsteed supplied some data already processed, but his assistants made mistakes. Newton was not impressed: 'I want not your calculations but your observations only … If you like this proposal, then pray send me first your observations for the year 1692 & I will get them calculated & send you a copy of the calculated places. But if you like it not, then I desire you would propose some other practicable method of supplying me with observations, or else let me know plainly that I must be content to lose all the time & pains I have hitherto taken about the Moon's theory.' Flamsteed was hurt, writing in his diary that Newton was 'hasty, artificial, unkind, arrogant'. It was the start of a long feud.

The star catalogue

The fundamental point at issue was a difference of opinions over the status of the star catalogue that Flamsteed was compiling. The Astronomer Royal saw it as his property and was determined that it not be published until it was complete. As far as Newton was concerned, Flamsteed's job was to give him what he wanted, and in 1705 he forced the issue. Through his influence at court, Newton encouraged Prince George, husband of Queen Anne, to commission Flamsteed to produce a *Historia Coelestis Britannica* ('A British History of the Heavens').

◀ **Newton's telescope.** *To illustrate his theory of optics, Newton built a new type of telescope, which used a mirror rather than a lens to gather light. But he was not an astronomer, and relied on Flamsteed for data.*

Flamsteed employed delaying tactics, but in 1711 was 'afresh disturbed', when Newton arranged for an inspection of the Royal Observatory. Later, Newton and Flamsteed tussled over the observatory's equipment: '… the President (Sir Isaac Newton) of the Royal Society … had formed a plot to make my instruments theirs … I had resolved aforehand his knavish talk should not move me; showed him that all the instruments in the Observatory were my own … This nettled him.'

▲ **Pugilistic philosophers** *A 19th-century cartoon showing Newton squaring up to Flamsteed and calling him a 'Puppy', an insult taken from Flamsteed's own testimony.*

WAS NEWTON PSYCHOTIC?

Newton may have been the most disputatious scientist in history. Not only did he hate to be disputed or opposed, he was also arrogant, dismissive, hostile, manipulative, vicious and able to hold a grudge for decades. Physician and antiquarian William Stukeley befriended Newton in his last years, diligently gathered biographical material about him and had only good words to offer. Yet he reported that when he made the minor error of applying for the post of Secretary to the Royal Society without first asking Newton, who was President, 'Sir Isaac showed a coolness towards me for two or three years.'

Newton displayed other signs of a troubled psyche. In 1693, he suffered a breakdown of some sort, penning disturbed, even deranged letters to friends, leading to speculation that he had suffered from mercury poisoning. Yet even as a child he was troubled, once threatening to burn down his mother and stepfather's house about them. The likely root of his dysfunctional psychology was childhood abandonment. His father had died before he was born and when he was three his mother remarried and moved in with her new husband, leaving Isaac behind in his grandmother's house. In addition, his biography shows signs of struggles with sexuality and maybe even of autistic spectrum disorders.

NEWTON vs HOOKE

Newton's first encounter with the wider intellectual world came in the 1670s, when a reflecting telescope he had made (see p178) was presented to the Fellows of the Royal Society (RS), followed by papers recounting experiments on light and colour that he had performed. 'I am purposing them,' he wrote in 1672, 'to be considered of & examined, an account of a philosophical discovery which induced me to the making of the said telescope, & which I doubt not but will prove much more grateful then the communication of that instrument.'

Newton's optical experiments were brilliant and powerful, but he was trespassing on territory that Robert Hooke, the influential Curator of Experiments at the RS, had made his own, and he resented the upstart and some of his conclusions. 'I have perused the excellent discourse of Mr Newton,' he responded, 'yet as to his hypothesis of solving the phenomenon of colours thereby I confess I cannot yet see any undeniable argument to convince me of the certainty thereof.'

Newton immediately lost his cool. Hooke's attack was 'a bare denial without assigning a reason', he complained to the Secretary of the RS, continuing: 'I was a little troubled to find a person so much concerned for an Hypothesis, from whom in particular I most expected an unconcerned & indifferent examination .... Mr Hooke thinks himself concerned to reprehend me ... But he knows well that it is not for one man to prescribe Rules to the studies of another, especially not without understanding the grounds on which he proceeds.'

Their feud rumbled on. In 1676, Newton submitted another paper on optics; once again, Hooke raised objections. In reply, Newton claimed that his experiments 'destroy all [Hooke] has said,' and snapped, 'I suppose he will allow me to make use of what I took the pains to find out.' A letter to Hooke a few weeks later included the now famous line: 'What Descartes did was a good step. You have added much several ways ... If I have seen further it is

◀ **Hooke's flea.** *In his landmark 1665 book* Micrographia, *Hooke opened up new worlds of the microscopic. The book was a major influence on Newton.*

by standing on the shoulders of giants.' The phrase was not original to Newton, but many commentators read the apparent flattery as a deliberate slight on Hooke, who was far from gigantic. In his *Brief Lives*, John Aubrey described Hooke as 'but of middling stature, something crooked, pale faced, and his face but little below, but his head is large'.

After their optical disputes, the pair tangled again over gravity and orbital mechanics. Newton, who had moved beyond Hooke in terms of his method and abilities, can generally be understood to have bested his enemy on most counts (although on one major point of dispute, the question of whether light is made up of waves or particles, both men were ultimately proved right). However, his loathing of Hooke meant that Newton stayed away from the Royal Society until the curator was dead, and he made sure to delete all mentions of Hooke from his works.

'honest Sir Isaac Newton (to use his own words) would have all things in his own power, to spoil or sink them'

JOHN FLAMSTEED, ASTRONOMER ROYAL, 1717

Tempers escalated, Newton writing angrily: '[if] you propose anything else or make any excuses or unnecessary delays it will be taken for an indirect refusal to comply with Her Majesty's order.' On 26 October, Flamsteed was summoned to appear before the council of the Royal Society, recounting what happened in his journal: '… all he said was in a rage: he called me many hard names; puppy was the most innocent of them … I only desired to keep his temper, restrain his passion, and thanked him as often as he gave me ill names.'

The following year, Newton succeeded in forcing the publication of the *Historia*, and the year after he was able to include the lunar theory in the second edition of the *Principia*. But Flamsteed had the last laugh. Queen Anne died in 1715 and Newton lost influence at court. Flamsteed was allowed to burn all the copies of the *Historia Coelestis* he could find and produce his own edition, finally published posthumously in 1725, although his venomous preface, accusing Newton of 'disingenuous and malicious practices', was suppressed for decades.

NEWTON

vs

LEIBNIZ

FEUDING PARTIES
Isaac Newton, (1642–1727)
– scientist, President of Royal
Society, inventor of calculus
vs
Gottfried Wilhelm von Leibniz
(1646–1716) – philosopher,
lawyer, diplomat, mathematician,
scientist

DATE
1684–1727

CAUSE OF FEUD
Priority in the invention of
calculus

Newton's worthiest foe was the German polymath Leibniz, a remarkable intellect described by Frederick the Great of Prussia as 'a whole academy in himself'. In 1673, Leibniz had visited London to present his calculating machine and forge connections with English intellectuals. He was elected a Fellow of the Royal Society (RS), and met the publisher John Collins, who served as a kind of clearing house for correspondence between philosophers and new developments in mathematics.

'The labour of thought'

Leibniz then returned to Paris and spent the next two years making remarkable advances, including the development of calculus (*see* box, page 185). There was only one problem: Newton had already invented this technique, which he called his 'method of fluxions', at least six years earlier. Leibniz's version was superior in point of view of its symbolic notation, which made it much easier to use, for as he observed: 'In symbols one observes an advantage in discovery which is greatest when they express the exact nature of a thing briefly and, as it were, picture it; then indeed the labour of thought is wonderfully diminished.'

Leibniz kept in touch with Collins, who realized that the German had duplicated Newton's discovery and foresaw potential controversy. The Englishman was urged to assert priority, but restricted himself to sending two letters to Leibniz in 1676 alluding to his breakthrough in cryptic tones (a kind of patent): 'I cannot proceed with the explanation of

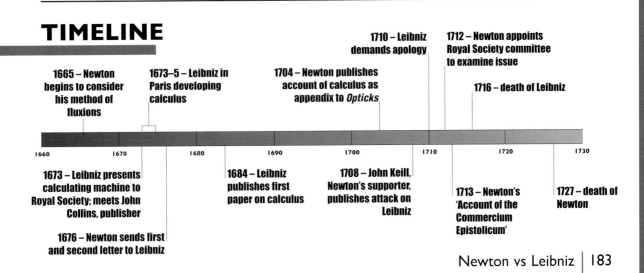

TIMELINE

1710 – Leibniz demands apology

1712 – Newton appoints Royal Society committee to examine issue

1665 – Newton begins to consider his method of fluxions

1673–5 – Leibniz in Paris developing calculus

1704 – Newton publishes account of calculus as appendix to *Opticks*

1716 – death of Leibniz

1660   1670   1680   1690   1700   1710   1720   1730

1673 – Leibniz presents calculating machine to Royal Society; meets John Collins, publisher

1684 – Leibniz publishes first paper on calculus

1708 – John Keill, Newton's supporter, publishes attack on Leibniz

1713 – Newton's 'Account of the Commercium Epistolicum'

1727 – death of Newton

1676 – Newton sends first and second letter to Leibniz

'Second inventors count for nothing'

SIR ISAAC NEWTON, 1713

the fluxions now, I have preferred to conceal it thus: 6accdae13eff7i3l9n4o4qrr4s8t12vx.' The code gave the numbers of letters in each word of a Latin phrase, which in turn translated to: 'Given an equation involving any number of fluent quantities, to find the fluxions: and vice versa.' In a covering note Newton wrote: 'I hope this will so far satisfy M. Leibniz that it will not be necessary for me to write any more ... it proves an unwelcome interruption to me to be at this time put upon considering these things.'

Newton probably thought the matter settled, but Leibniz had paid little attention to his claims and in 1684 blithely published an account of his new system, which he called calculus. There was no mention of Newton, and when it was suggested to him that there might be a problem, he naively brushed aside the concern: 'I acknowledge that Mr Newton already had the principles ... but one does not come upon all the results at one time; one man makes one contribution, another man another.'

Newton's lost children

Newton was presumably furious, but engaged as he was in writing the *Principia* and other concerns he let others fight his battle for him. These proxies, primarily John Wallis, Fatio de Duillier and John Keill, so exasperated Leibniz over the next few years that he came to call them Newton's *enfants perdu* ('lost children'). Wallis was particularly

▶ **Leibniz's Reckoner.** *A reconstruction of the calculating device Leibniz presented to the Royal Society in 1673, known as the 'Stepped Reckoner' because of its ingenious stepped drum mechanism.*

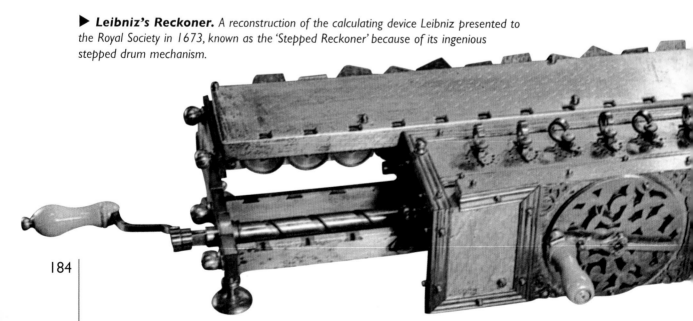

WHAT IS CALCULUS?

Calculus is a mathematical system for calculating the gradients of curved lines and the areas under them, feats beyond the capabilities of the ancients (the ancient Greeks, for instance, struggled for centuries with the problem of calculating from the dimensions of an amphora or barrel with curved sides exactly how much wine it would hold). Calculus was also necessary for solving problems of motion, in which the speed of moving points or objects (such as planets) changes over time. The key breakthrough of both Newton and Leibniz was to develop a way of working with 'infinitesmals' – quantities tending to zero but which are not actually zero. The priority dispute was bad news for British science. National pride meant that for well over a century British mathematicians were saddled with Newton's method of fluxions, which was much more difficult to work with than Leibniz's calculus.

concerned with national pride, complaining to Newton in 1695 that 'You are not so kind to your Reputation (& that of the Nation) as you might be when you let things of worth lie by you so long, till others carry away the Reputation that is due to you.'

The two camps traded blows. Fatio de Duillier stirred the pot by publishing an attack on Leibniz: 'Neither the silence of the more modest Newton, nor the remitting exertions of Leibniz to claim on every occasion the invention of the calculus for himself, will deceive anyone.' In 1705 an anonymous review of *Opticks*, widely known to have been penned by Leibniz, provocatively compared Newton to notorious plagiarist Honoré Fabri, and in 1708 Keill published another thinly veiled attack on Leibniz, which provoked an angry response demanding an apology.

In 1712, Newton used his position as President of the Royal Society to convene a committee to examine the dispute. Although supposedly impartial, it was packed with placemen and Newton wrote the report, following it up with a vitriolic and personal commentary on proceedings that amounted to an attempted character assassination of Leibniz. Leibniz died in 1716 and Newton displayed his enduring malice by boasting many years later that he had 'broke Leibniz's heart with his reply to him'.

WALLACE

VS

HAMPDEN

FEUDING PARTIES
Alfred Russel Wallace
(1823–1913) – biogeographer,
co-discoverer of the theory of
natural selection, spiritualist,
globuralist
vs
John Hampden (died 1891) and
William Carpenter (1830–96)
– planists

DATE
1870–91

CAUSE OF FEUD
The roundness of the Earth

Several passages in the Bible appear to support the idea that the Earth is flat, talking of the 'four corners' of the world and the 'pillars of the Earth'. This chimes with common sense: the Earth appears to be flat to anyone standing on its surface, and the notion that it is in fact a giant globe hurtling through space while revolving at tremendous speed seems contradicted by the evidence of everyday life. For instance, if you throw a ball straight up in the air it will come straight back down and land on your head; how can this be if the Earth has revolved several metres in the time the ball has been in flight?

Parallax and the Old Bedford Level

The answers to some of these questions have been known since ancient times; others were provided by the likes of Galileo and Newton. The Earth only appears flat to terrestrial observers because its curvature is too slight to be obvious to the naked eye unless very high off the ground. Balls thrown aloft come straight back down because they and everything else on the surface partake of the Earth's radial motion. There are many other proofs of the curvature of the Earth's surface, the best known being the 'hull-down' phenomenon in which the masts and sails of ships disappearing over the horizon remain visible after the hull has vanished from sight. Flat-Earthers or 'planists' dispute this interpretation. One of the most energetic 19th-century planists, Samuel Rowbotham, who went by the pseudonym 'Parallax', set out to disprove it on an unusually straight, flat canal: the Old Bedford Level in Cambridgeshire in England, where between the bridge at Welney and the one at Old Bedford there was an uninterrupted stretch of six miles (ten kilometres).

According to conventional wisdom, the curvature of the Earth should make it impossible for someone at water level at the Old Bedford end to view an object at water level beneath the Welney bridge. Yet Rowbotham spent years demonstrating otherwise to parties of interested visitors, detailing the results of his experiments in his book

▶ *Alfred Russel Wallace. From a relatively humble background, Wallace trained as a surveyor before becoming a naturalist and launching the expeditions that brought him scientific renown.*

> ## 'I must decline to spend any more time in refuting arguments founded on total ignorance alike of facts and of geometrical principles.'
>
> WALLACE TO CARPENTER, 1870

Zetetic Astronomy ('*zetetic*' derived from the Greek for 'I find out for myself'). A fan of zetetics and follower of Parallax, John Hampden of Swindon issued a challenge to globularists (those who maintain the Earth is a globe) in 1870, wagering £500 (a considerable sum at the time) against anyone who would take part in a new experiment on the Old Bedford Level. The wager was accepted by Alfred Russel Wallace, co-discoverer, along with Charles Darwin, of the theory of evolution by natural selection. Wallace had lost money in unwise investments and perhaps assumed that with his background in surveying (he had trained in the trade as a young man) he could easily disprove the planist contentions and earn some money in the process.

A matter of perspective

Hampden and Wallace each chose a man to act as a referee, the planist choosing William Carpenter, a fellow flat-Earth enthusiast. Convening at the canal, they devised an experiment with a signal on the face of the Old Bedford bridge, 13 feet 4 inches (4 metres) above the water, and a disc on a pole at the same height halfway along the stretch of water. Using a surveyor's telescope with spirit level perched on the parapet of the bridge at Welney, also 13 feet 4 inches (4 metres) above the water, the antagonists viewed the canal. Wallace saw, as he expected, that the central disc appeared to be 5 feet (1.5 metres) higher than the more distant signal, just as predicted by the theory of curvature. Yet Hampden and Carpenter were triumphant: what they saw was that the crosshair in the viewfinder was above the central disc, which they interpreted as an effect of perspective that proved they were right.

FLAT EARTH THEORIES

The ancient Greek philosopher Eratosthenes (c.276–195 BCE) noted that on the summer solstice in Alexandria, when the Sun was directly overhead so that vertical poles cast no shadows, obelisks in Syene directly to the south *did* cast shadows. From this, he inferred that the surface of the Earth was curved and, knowing the distance from Alexandria to Syene, he used geometry to calculate the radius and circumference of the Earth, arriving at a figure of 250,000 stadia, or 24,662 miles (39,690 kilometres), less than 300 miles (483 kilometres) off the value adduced by modern methods. Yet there were plenty who maintained belief in a flat Earth; and, thanks to the 6th-century *Christian Topography* of Cosmos Indicopleustes, the Church came round to the idea that the spherical doctrine was pagan. The Bible mentioned the 'four corners of the Earth', and Cosmos suggested that the vault of heaven was fastened to these forming a 'vast canopy' over the Earth.

This was the beginning of a long and surprisingly resilient idea, which found its fullest flowering in the 19th century. Planists, such as Parallax, were expert contrarians, and their ranks spawned many eccentric characters. In 1905, the enthusiastic zetetic scholar Lady Blount led her own expedition to the Old Bedford Level, employing a cameraman to take photos from very near the water level at the Welney bridge end. The cameraman was able to obtain a clear photo of the whole of a large calico sheet draped over the edge of the Bedford bridge, right down to the water level, including its reflection in the water. This was problematic for globularists – a few years earlier, the popular Victorian astronomy writer Richard Proctor had asserted: 'If with the eye a few inches above the surface of the Bedford Canal, an object close to the water, six miles [9.7 kilometres] distant from the observer, can be seen, then manifestly there would be something wrong with the accepted theory.'

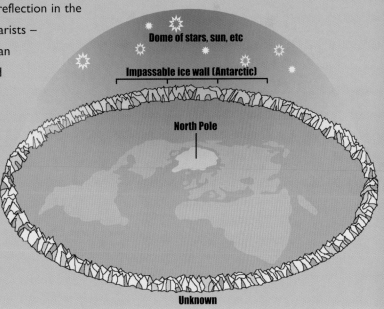

▶ *Flat Earth cosmography*

Wallace insisted the placing of the crosshair was irrelevant, and the wager descended into bad-tempered dispute. A referee was called in to adjudicate and, after consulting with instrument makers, decided in favour of Wallace and paid over the money. Hampden and Carpenter were furious, the latter firing off a letter to *The Field* journal disputing Wallace's claims and methods, to which the scientist replied testily: 'the fallacies and misstatements with which it abounds … may perhaps confuse and mislead some of your readers who are not very conversant with practical geodesy.' He accused Carpenter of 'defining a "straight line" in a manner totally new', and of being 'utterly confused and misleading' about 'a question of elementary geometry about which there can be no dispute'. He roundly dismissed all of Carpenter's objections: 'No. 4 is an entire delusion. No. 5 is an assertion destitute of proof. Nos. 6 and 7 are verbal quibbles with which I have nothing to do. Nos. 8, 9, and 10 rest on [a] fallacy … which is a pure figment of Mr C.'s brain … Nos. 11, 12, and 13 are misconceptions.' 'All his wordy argumentation is utterly valueless,' he concluded, suggesting that if Carpenter wanted to persist in his delusions, he should 'prove [his] elementary point by experiment and diagrams, and thus found a totally new and hitherto unimagined geometry'.

WALLACE AND SPIRITUALISM

Wallace's brilliant work on zoology and botany in South-east Asia, his co-discovery of natural selection and his expert explication and defence of evolutionary theory won him admirers across the world of science, yet he was no stranger to scientific controversy. Having become interested in spiritualism – the phenomena of séances, mediums and contact with the dead – in the 1860s, he went on to become one its most fervent and high-profile advocates. Wallace believed uncritically in the entire range of spiritualist phenomena, defending transparently bogus mediums who had been caught in acts of outrageous fraud. Even those sympathetic to spiritualism had harsh words for him. Frederick Myers of the Society for Psychical Research complained of his 'resolute credulity' and lack of even 'a trace of scientific instinct or training'. Physicist Oliver Lodge described him as 'a crude, simple soul, easily influenced, open to every novelty and argument'. When Wallace began to consider 'heretical' notions of design in human moral and mental evolution, Darwin accused him of having 'murdered too completely your own and my child'.

▲ **On the level.** *A view of the Bedford level, a region of the fenlands of eastern England remarkable for its flatness, drained by long, straight, level canals – the perfect proving ground for planist theories?*

To Hampden, he had this to say: 'Mr Hampden, in his letter to me, continually appeals to "public opinion" as being against the fairness of [the] verdict. It has, however, now clearly spoken through your widely-circulated columns, and … he would do well, as a man of honour and of sense, to bow to its decision.' Unfortunately, Hampden seemed to possess little of either honour or sense, and spent the next 21 years hounding and abusing Wallace, in the courts, in print and with threatening poison-pen letters to him and his family.

A typical example sent to Wallace's wife read: 'Madam, If your infernal thief of a husband is brought home some day on a hurdle, with every bone in his head smashed to a pulp, you will know the reason. Do you tell him from me he is a lying infernal thief, and as sure as his name is Wallace he never dies in his bed. You must be a miserable wretch to be obliged to live with a convicted felon. Do not think or let him think I have done with him. John Hampden.' Despite lengthy court proceedings and spells in jail, Hampden continued his vendetta until the end of his life, nearly bankrupting Wallace with legal costs.

WAGERS IN SCIENCE

As this book amply demonstrates, disagreement is a driving force in science. At the core of the scientific method is experimental proof or disproof of a hypothesis but until that experimental judgement is handed down the hypothesis is a matter of conjecture, speculation … opinion. Like people in any other walk of life, scientists often find themselves holding contrasting opinions. What makes science different is the expectation that questions will be clearly answered, that speculations will be proved wrong or right, that opinion will give way to certainty, and this makes science fertile ground for wagers – where a definite resolution is possible, a wager can be definitively settled.

'To encourage the inquiry'

To make or accept a wager, however, a scientist must have self-confidence, and some of the earliest known scientific wagers arguably reflect egotism more than courage. In 1600, the astronomer Johannes Kepler bet rival Christian Longomontanus that he could derive the formula for the orbit of Mars in just eight days. Kepler recorded the existence of the wager but not the stakes, so it is not known what he lost when it actually took him five years. His findings helped to lay the foundations for the Newtonian system that was to come some 80 years later, laid out in the pages of Newton's *Principia*.

Coincidentally, the *Principia* itself partly owed its existence to a wager. On a Wednesday in January 1684, Sir Christopher Wren, Robert Hooke and Edmund Halley were sitting in a coffee house discussing matters philosophical. Kepler had discovered that the orbits of the planets were governed by an inverse square law and all three of those present had, like Newton much earlier, guessed that this same law might govern the attractive power keeping them in their orbits. Halley ventured to suggest as much, prompting Hooke to 'affirm that upon that principle all the Laws of the celestial motions were to be demonstrated, and that he himself had done it'. Perhaps mindful that Hooke was notorious for hollow boasts, 'Sir Christopher to encourage the Inquiry said that he would give Mr Hooke or me 2 months time to bring him a convincing demonstration thereof, and besides the honour, he of us that did it, should have from him a present of a book of 40 shillings.'

Hooke claimed that he would, in due course, 'make it publick', but Halley recalled: 'I remember Sir Christopher was little satisfied that he could do it, and though Mr Hook then promised to show it him, I do not yet find that in that particular he has been as good as his word.' Neither man won Wren's bet and after waiting two months and more for Hooke's proof, Halley set off to Cambridge to seek answers from Newton, a visit that led Newton to start writing the *Principia*.

Edmund Halley, here seen in old age. As a younger man he was an acolyte of Newton's. Although considered charming and affable, he managed to arouse the ire of John Flamsteed, who despised his rumoured atheism.

Sir Christopher Wren, depicted leaning on the plans for the new cathedral of St Paul's. This portrait was painted in 1711, the year after Wren's masterpiece, replacing the church that had burned down in 1666, was finally completed.

'A friendly wager'

Kepler and Wren kicked off an ongoing trend in their fields, for it seems that scientific wagers are particularly common in physics and astronomy. Kevin Kelly, co-founder of Long Bets, a website that provides a forum for scientific wagers, explains: 'Physics and cosmology are such quantitative fields that they readily lend themselves to wagers. And since it can take decades to build the high-priced gear needed to prove a theory, long-term betting is a way of dispensing credit for early insight.'

Prominent physicist punters include Richard Feynman and Stephen Hawking. In 1957, Feynman lost a bet on whether left–right symmetry is maintained in subatomic reactions (it isn't), and then lost another bet two years later when he issued a US\$1,000 challenge that nobody could construct a motor smaller than ⅟₆₄th of an inch (0.4 millimetres). Feynman had hoped the bet might stimulate development of a radical new technology and/or material, and was disappointed when scientific instrument maker Bill McLellan used existing technology to achieve the feat.

Hawking meanwhile, has engaged in at least three high-profile wagers. In 1974, he was involved in what his biographer Kristine Larsen calls 'A friendly wager that was to become the stuff of legend' with physicist Kip Thorne, over whether Cygnus X-1 would turn out to contain a black hole. The pair drew up a covenant stating: 'Whereas Stephen Hawking has such a large investment in General Relativity and Black Holes and desires an insurance policy, and whereas Kip Thorne likes to live dangerously without an insurance policy

PHILOSOPHIÆ
NATURALIS
PRINCIPIA
MATHEMATICA.

Autore JS. NEWTON, Trin. Coll. Cantab. Soc. Matheseos Professore Lucasiano, & Societatis Regalis Sodali.

IMPRIMATUR·
S. PEPYS, Reg. Soc. PRÆSES.
Julii 5. 1686.

LONDINI,

Jussu Societatis Regiæ ac Typis Josephi Streater. Prostat apud plures Bibliopolas. Anno MDCLXXXVII.

Frontispiece of Newton's Principia. Generally regarded as the most important and greatest scientific work ever published, the Principia, or Mathematical Principles of Natural Philosophy, arguably owed its existence to a wager, since Newton was prompted to write it by the visit of Halley, who was seeking to settle a bet.

… Therefore be it resolved that Stephen Hawking bets 1 year's subscription to 'Penthouse' as against Kip Thorne's wager of a 4-year subscription to 'Private Eye', that Cygnus X-1 does not contain a black hole of mass above the Chandrasekhar limit.' Hawking was using the bet as a form of hedge, and had offered Thorne odds of 4-to-1 on because 'we were 80 per cent certain that Cygnus was a black hole'. It was and he duly paid up.

In 1997, Hawking and Thorne accepted a wager from fellow physicist John Preskill that: 'When an initial pure quantum state undergoes gravitational collapse to form a black hole, the final state at the end of black hole evaporation will always be a pure quantum state.' Hawking subsequently accepted defeat again, but was still game enough to accept another wager in 2000, this time betting another physicist, Gordon Kane, US$100 that the hypothesized particle, the Higgs Boson, would be discovered.

Wagers in the world of physics are so popular that some labs run notorious betting books. The Bell Laboratories in New Jersey, USA, ran one for decades (until it disappeared in 1990, possibly stolen by a welcher), and more recently the Stanford Linear Accelerator Center Theory Group has been keeping an Official Record of Wagers that runs to 60 bets.

Betting on disaster

Physicists are not the only ones making wagers. As environmentalists and those in related fields have become more vocal about the calamitous future of the Earth, so sceptics have also become more vehement. In several cases, one side has challenged the other to put its money where its mouth is. The most famous example was Paul Ehrlich's 1980 wager with economist Julian Simon on the price of mineral commodities. Ehrlich was a popular and influential predictor of environmental doom because of population explosion, and one of his predictions was that increasing pressure on world resources would cause commodities prices to rocket. He and Simon wagered on what a basket of US$200-worth of each of five minerals would be worth in ten years. By 1990, the value of the basket had almost halved; Ehrlich lost the best. Kelly points out that 'nothing that Simon ever wrote had as much impact on the course of culture as his wager with Ehrlich. That single, relatively small bet transformed the environmental movement by casting doubt on the notion of resource scarcity.'

Doom-mongers remain undaunted, however. In 2005, British climate expert James Annan, after failing to interest climate-change sceptic Richard Lindzen, who demanded odds of 50–1, succeeded in getting two Russian scientists to take a US$10,000 bet that average global temperatures between 2012 to 2017 would be higher than between 1998 and 2003. 'There isn't much money in climate science,' noted Annan, 'and I'm still looking for that gold watch at retirement. A pay-off would be a nice top-up to my pension.'

TESLA
vs
EDISON

FEUDING PARTIES
Nikola Tesla (1856–1943)
– inventor;
George Westinghouse (1846–1914)
– inventor and businessman
vs
Thomas Alva Edison (1847–1931)
– inventor and businessman;
Harold P. Brown (1869–1932)
– electrical engineer, inventor of
the electric chair

DATE
1884–1903

CAUSE OF FEUD
War of the Currents

The invention of technology to generate, distribute and use electricity led to what has been termed the Second Industrial Revolution, and ushered in the Power Age. Today, the legendary American inventor Thomas Edison gets most of the credit for this techno-industrial revolution but the plaudits more properly belong to the Serbian inventor Nikola Tesla and his American backer, George Westinghouse, himself an inventor of note.

In the early 1880s, Edison made headlines and a fortune by introducing electricity for everyday domestic and retail use, generating it at small power plants, such as Pearl Street Station in New York, sending it short distances down thick copper wires to light streets, shops and homes in the neighbourhood. Edison had built his system around direct current (DC), but in 1888 a new and potentially far superior technology came into the market, built around alternating current (AC). Edison,

▼ **Edison's lab.** *A laboratory in Fort Myers, Florida, which Edison used towards the end of his life, when he was searching for a synthetic alternative to rubber.*

> ## 'The alternating current can be described by no adjective less forceful than damnable.'
>
> HAROLD BROWN, 1888

determined to avoid the ruinous cost of licensing the rival patents and completely retooling his systems, launched a vicious campaign to smear the upstart technology. As the two systems fought to become the universal standard and reap the huge profits on offer, their rivalry descended into a bitter feud that became known as the War of the Currents. Tom McNichol, author of *AC/DC: The savage tale of the first standards war*, describes how, 'In the AC/DC battle, the worst aspects of human nature somehow got caught up in the wires, a silent, deadly flow of arrogance, vanity, and cruelty.'

▼ **Master of lightning.** *Tesla seated in his laboratory while massive arcs of electricity crackle around him. His ability to harness immense voltages appeared almost magical.*

TESLA THE ELECTRIC SORCERER

Tesla described Edison as an inventor and himself as a discoverer. During his lifetime his 'discoveries' helped change the world and make him a household name. Yet today he is primarily known for his associations with weird science and conspiracy theories. After selling his AC patents to Westinghouse and helping the company build much of its pioneering infrastructure (including in 1893 massive turbines to harness the power of Niagara Falls), Tesla turned his attention to researching the strange world of very high-voltage, high-frequency electricity. His mastery of the technology meant he was able to pass huge currents through his body without ill effect but producing spectacular effects, including fizzing blue auras and the ability to power fluorescent tubes (which he had also invented) by holding them. He would demonstrate these abilities in stage shows that earned him the reputation of being an 'Electric Sorcerer'.

Later, Tesla tried to develop wireless transmission of electricity, but his visionary plans ran out of money and he retreated into eccentricity and outlandish claims about new technology, such as death rays, earthquake weapons and weather control systems. When he died in 1943, there was a strange tussle over his papers. Officially, they belonged to his nephew, the Yugoslav ambassador, but the FBI was called in and the papers were seized by the US Government. Most were given back, but some are still classified today, leading to extensive speculation about Tesla technologies being developed by secret government research projects and doomsday cults.

'American humour'

The brains behind the AC technology was an eccentric young man from Europe, Nikola Tesla, who had come to America to work for Edison bearing a letter from the great inventor's partner, Charles Batchelor, which read: 'I know two great men and you are one of them: the other is this young man.' Tesla was nursing brilliant ideas for overcoming technical limitations that had previously made the use of AC electricity impractical but to his dismay he found that Edison was too invested in DC technology to be interested. Relations between the men soured for good when Edison reneged on a verbal contract to pay Tesla US$50,000 if he could improve the design of his company's turbines. Tesla completed the task only to be told the offer had been a joke: 'Tesla, you don't understand our American humour.'

Edison had made a bad mistake. Tesla went away and invented a series of revolutionary technologies for generating, distributing and using AC power. The new system had the advantage that it could be transmitted at very high voltages, allowing it to travel much farther, much more efficiently and through much thinner wires than DC power. At the point of distribution the dangerously high voltages could be transformed to much lower ones, suitable for domestic use. Thinner wires meant less copper, while long-distance transmission meant fewer stations: the AC system promised to be much cheaper. When George Westinghouse bought up Tesla's patents and started aggressively selling AC, Edison was in trouble and he knew it.

The wire's fatal grasp

Efficiency and cost were not the only weapons in this war, however, and Edison was a master of marketing and spin. Electricity was a new and unknown quantity to the man and woman in the street, and public perception was primed for scare stories that would play on their darkest fears. The scramble for electrification had resulted in many poorly installed lines, which in turn generated shocking headlines such as 'THE WIRE'S FATAL GRASP' and 'AGAIN A CORPSE IN THE WIRES'.

A new name now entered the picture. Harold Brown had started life as a salesman for Edison's electrical pens before setting himself up as an electrical engineer; he would lead the campaign against AC, calling it a 'damnable death current' while insisting DC was 'completely harmless'. Attacked in print by supporters of AC, Brown hatched a plan: 'I must show from their own current and its effect upon life as compared with continuous currents that my statements are true. Words are of no avail against such accusations as theirs.'

Although Edison had no official links with Brown, the inventor quickly realized that Brown could be a useful stooge, and supplied him with equipment and credibility. The former salesman began a grisly programme of experiments and demonstrations

◄ *Damnable death current.* *A cartoon of 1891 reflecting contemporary fears about the dangerous new utility being rolled out to American cities. In the rush to electrify, some of the infrastructure was shoddily installed, leading to scare stories about accidents.*

in which he electrocuted dogs with first DC (which they survived) and then AC (which they did not). In mid–1888, Brown staged a demonstration in which he tortured a Newfoundland dog with bursts of DC current, before finishing it off with AC, to prove that the latter was more lethal. The meeting was broken up by an animal rights official and an angry crowd demanding that Brown submit to a human electrical duel. 'I wish this experiment had not been interrupted,' he protested. 'I have enough dogs to satisfy the most sceptical.'

In December 1888, Brown moved on to electrocuting calves and a horse, and the following year he designed the first electric chair for human executions, using an illegally purchased Westinghouse generator. Although his device was hailed as 'a highly scientific device for electrical executions' by the *New York World*, AC was actually the worst option for a quick, painless execution. On 6 August 1890, murderer William Kemmler became the first victim of the chair: it took two

▲ **Festival of Light.** *A portrait of the Electrical Building at the Chicago World's Fair of 1893, brightly lit at night thanks to the transforming power of electricity. After a fierce tussle to win the contract to provide light and power, the Fair proved to be the ultimate showcase for Westinghouse and his Tesla-invented technologies.*

TOPSY THE BAD ELEPHANT

The War of the Currents claimed its final victim in 1903 – an unfortunate elephant named Topsy. Topsy was a circus elephant that had been dragged around the United States for 28 years, but through bad care and abusive treatment became ill-tempered and dangerous. She killed two keepers in Texas, and another in Brooklyn when he tried to feed her a lit cigarette. It was decided she had to be put down, but the best mode of execution was unclear until Edison volunteered to supply equipment to electrocute Topsy. A crowd of over a thousand gathered at New York's Coney Island to see the show. Topsy was fed one pound (half a kilogram) of cyanide-laced carrots, and fitted with copper-lined sandals, before 6,600 volts were blasted through her body. 'There had been no sound and hardly a conscious movement of the body,' reported the *New York Times*, describing the event as 'a rather inglorious affair'. 'Bad Elephant Killed', announced the *Commercial Advertiser* for 5 January 1903, claiming that 'the big beast' had died 'without a trumpet or a groan'.

Edison filmed the whole sorry show and the footage toured the country, being shown to tens of thousands. In 2003, a museum on Coney Island unveiled a memorial to Topsy, designed by artists Gavin Heck and Lee Deigaard. 'She was considered a bad elephant, but she wasn't a bad girl,' Deigaard told the BBC, while Heck admitted he was 'struck by how a story so old could bring up so many issues and feelings for people'.

attempts to kill him and the execution descended into grim farce, described as 'an awful spectacle, far worse than hanging'. As Westinghouse remarked, 'They would have done better with an axe.'

The gruesome showmanship orchestrated by Edison was to no avail. Westinghouse's system was better and cheaper, and when he and General Electric (the newly formed company controlling Edison's patents) came to bid for the contract to light the upcoming Chicago World's Fair (aka the Columbian Exposition) Westinghouse was able to undercut the rival bid by half. On 1 May 1893, President Grover Cleveland switched on 100,000 light bulbs largely powered by AC, and the War of the Currents was won. GE finally admitted defeat in 1896 when they cross-licensed Westinghouse's patents.

HOYLE

vs

RYLE

FEUDING PARTIES
Sir Fred Hoyle (1915–2001)
– astrophysicist
vs
Sir Martin Ryle (1918–84)
– radio astronomer

DATE
1950s–2001

CAUSE OF FEUD
Steady state vs Big Bang

The face of astronomy in the UK for several decades after the Second World War was the charismatic astrophysicist Sir Fred Hoyle, who revolutionized both his field and the public perception of it. But Hoyle was an inveterate contrarian, with a habit of picking fights he could not win, which often pitted him against the mainstream of science.

After important research work on radar during the war, Hoyle turned his attention to the exciting field of stellar physics. It had been shown that the stars ran on fusion, a nuclear reaction in which hydrogen atoms are fused to create helium, but the origin of the rest of the elements remained a mystery. Together with the American William Fowler, Hoyle showed exactly how the other elements are created in stars, coining the famous phrase 'we are all made of stardust'. Years later, Fowler would be awarded the Nobel prize for his work; Hoyle was snubbed. According to astronomer Patrick Moore, it was 'a flagrant injustice' and Hoyle 'made no secret of his annoyance'.

In the beginning

Since the big question 'Where did we come from?' had been answered, Hoyle turned his attention to another: 'Where did it all begin?' It was known from the 'red shift' of light from distant galaxies that the universe was expanding (due to the Doppler effect, the wavelength of light from a receding object is longer and thus redder), and this prompted astronomers to assume it had started off very small, which in turn suggested there was a

TIMELINE

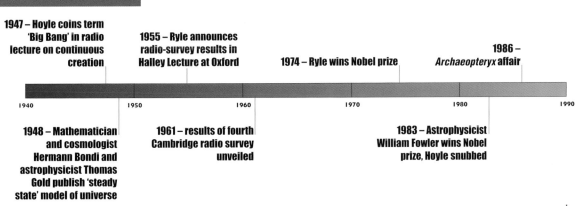

1947 – Hoyle coins term 'Big Bang' in radio lecture on continuous creation

1955 – Ryle announces radio-survey results in Halley Lecture at Oxford

1974 – Ryle wins Nobel prize

1986 – *Archaeopteryx* affair

1940   1950   1960   1970   1980   1990

1948 – Mathematician and cosmologist Hermann Bondi and astrophysicist Thomas Gold publish 'steady state' model of universe

1961 – results of fourth Cambridge radio survey unveiled

1983 – Astrophysicist William Fowler wins Nobel prize, Hoyle snubbed

'I have little hesitation in saying that a sickly pall now hangs over the big bang theory.'

FRED HOYLE, *THE INTELLIGENT UNIVERSE*, 1983

start point, a beginning before which the universe did not exist. Hoyle was dissatisfied with this explanation, gravitating towards the suggestion made in 1946 by his close friend and colleague Thomas Gold that perhaps the universe had always existed and always would. Over the next two years, Hoyle, Gold and fellow astrophysicist Hermann Bondi worked out a model that the latter two described as the 'steady state' model. The universe was expanding but that was because as old galaxies died out, new ones were born to replace them, in a process Hoyle termed 'continuous creation'.

A year earlier, Hoyle had begun his radio broadcasts, bringing the excitement of astrophysics and its attempts to answer the great questions to a popular audience. The avuncular manner of his 'fireside chats', together with his grand themes, proved an unexpected hit. It was in the first of these broadcasts that Hoyle coined the term 'big bang', which he used in derisive manner to describe the theory he opposed.

Although Hoyle was becoming a household name, he was not universally popular with his peers. His nemesis would prove to be Martin Ryle, a radio astronomer (one who uses telescopes that pick up radio waves instead of light waves) from a patrician, academic background very much at odds with Hoyle's provincial origins. He did not suffer fools gladly; according to Sir Bernard Lovell, 'Ryle's extraordinary inventiveness and immediate scientific insight … often led him to be intolerant of those not similarly blessed.'

Ryle had also worked on radar during the war, and afterwards had helped perfect the art of radio telescopy. He supported the Big Bang, or inflationary, model and his surveys of radio sources in other galaxies around the universe promised to settle the issue. In Hoyle's continuous creation model, galaxies (and therefore radio sources) should be evenly distributed in time and therefore around the universe. The Big Bang model, with its single moment of creation, predicted that most radio sources would be old and far away.

◀ **Radio days.** *Recently built radio telescope (part of the One Mile Telescope array) at the Mullard Observatory near Cambridge in 1965. Martin Ryle, co-founder of the Observatory, is the man in the centre foreground.*

Collapsing theory

In 1955, Ryle announced the results of his first 'Cambridge survey' of radio sources at the Halley Lecture in Oxford, revealing with something of a flourish that the results conclusively showed a great preponderance of faint, remote, old radio sources. The steady state model could not be correct. Hoyle was not impressed with the way that Ryle had announced his apparent victory, and was therefore delighted when the

▶ **Stellar nursery.** *A view from the Hubble Telescope showing young stars in the Carina spiral arm of our galaxy. Much older and more distant objects, such as pulsars and quasars, hold the key to the mystery of the origins of the universe.*

BIG BANG BUST-UP

While Ryle's work on radio astronomy may have strengthened the case for the Big Bang model, it is not without problems. In 1929, the American astronomer Edwin Hubble formulated an equation to describe the speed at which distant galaxies are receding from the Earth; in effect, the speed at which the universe is expanding. The outcome of this calculation tells us the likely fate of the universe – will it expand forever, or collapse in a 'big crunch' that is the reverse of the Big Bang? The formula features a quantity known as the constant of proportionality or the Hubble constant, effectively the speed in metres per second of a galaxy a distance of 1 megaparsec (mpc) away.

One of the most acrimonious feuds in physics was over the value of this constant. French astronomer Gérard de Vaucouleurs argued that its value was about 100, while Allan Sandage of the Carnegie Observatories in California held that it was about 50. The dispute became bad tempered; at a 1976 conference in Paris, for instance, 'they were duking it out in front of the audience', according to John Huchra of the Harvard–Smithsonian Centre for Astrophysics. Many in the field were unimpressed; Barry Madore of the California Institute of Technology described the feud as 'a battle not so much of the titans as of the grumpy old men'. Space telescope observations eventually evaluated the constant at about 70 (km/s)/Mpc, but Sandage refused to accept the results, telling the journal *Science*: 'There's still a controversy, and this isn't going to settle it.'

Cambridge survey turned out to have serious problems. Other astronomers could not replicate Ryle's findings, and it became apparent that his analysis was flawed, with systematic overestimation of the faint sources.

The dispute between Ryle and Hoyle became quite acrimonious; both men had spiky personalities. Dr Virginia Trimble, an astrophysicist at the University of California, says: 'I knew them both. Neither person was overwhelmingly easy to get along with.' Ryle and colleagues redid their analysis and corrected their errors, but their credibility was badly damaged, and the dispute rumbled on through three more Cambridge surveys. The media enjoyed the scientific scrap. In his 1994 memoir, *Home is Where the Wind Blows: Chapters from a Cosmologist's Life*, Hoyle recalled that after one prominent anti-Hoyle announcement, the national newspapers picked up on the story and 'for the next week, my children were ragged about it at school.'

▼ **Deep space explorers.** *An array of deep space radio telescopes point into space. These are designed to hunt for deep space phenomena, such as quasars and pulsars, which are radio sources that can indicate the age of the Universe.*

THE *ARCHAEOPTERYX* AFFAIR

Some of the most impressive pieces of evidence for the theory of evolution are the various fossils of *Archaeopteryx lithographica*, a fossil avian that shows clear evidence of both dinosaur and modern avian characteristics. The specimen held by the British Museum of Natural History (BMNH) is particularly fine, with visible impressions of feathers startlingly similar to those of modern birds. Opponents of evolution have targeted *Archaeopteryx*; one such was Fred Hoyle.

Hoyle rejected Darwinian evolution, claiming that life on Earth had started through 'seeding' from space rocks, that new mutations were constantly arriving from space and that all this was guided by a super-intelligent extraterrestrial civilization. Although Hoyle's 'panspermia' beliefs are no longer considered as far-fetched as when he first broached them in the 1970s (see pages 212–213), his views on evolution still seem outlandish. In 1986, Hoyle and his long-time collaborator Chandra Wickramasinghe published a book attacking the provenance of the BMNH *Archaeopteryx* specimen, suggesting that it was a fake: 'Our contention is that the feather impressions were forged onto a fossil of a flying reptile.' The impressions had been made, they suggested, by spreading a thin layer of concrete, made from the same limestone as the rest of the specimen, around the fossil, and pressing modern bird feathers into it.

Hoyle's arguments and evidence were given short shrift by experts. A key plank of his team's argument was analysis of photographs they had taken of the specimens, on which the BMNH commented damningly: 'the cursory examination and poor photographs of the authors of the [forgery accusation] articles bear no comparison with the close scrutiny and exacting standards of the Museum.' The palaeontologist Beverly Halstead, reviewing Hoyle and Wickramasinghe's book for *New Scientist*, did not mince his words: 'libellous nonsense … one of the most despicable pieces of writing it has been my misfortune ever to read'.

In 1961, the revised fourth Cambridge survey seemed to settle the matter in Ryle's favour. Although Hoyle attempted to rescue his theory by recasting it in modified form, the Big Bang model is now the generally accepted one (although it is still not free of problems). As late as 1999, Hoyle was still insisting it was bogus, describing it as 'a huge facade based on no real evidence at all'. 'The really interesting question,' he claimed, 'is why the world was so ready to believe in this story.'

FRINGE SCIENCE

Science is an investigation into the unknown, its field of enquiry the borderlands of knowledge. In this sense all science is fringe science. But there is a more specific meaning to the term, relating to ideas or projects that mainstream science has left behind, sidelined or tried to ignore altogether. This fringe is fertile territory for cranks, frauds, pseudo-scientists and, occasionally, visionary geniuses.

Occult forces

There is a case to be made that science was born out of fringe pursuits. Alchemists and magi in the Early Modern era sought to discover hidden laws and correspondences to manipulate and master the natural world – a useful definition of technology. The classic example of the esoteric giving birth to the scientific is Newton's conception of gravity. In the teeth of considerable opposition by those who thought his ideas too fantastical, Newton insisted that gravity was a force that acted at a distance with no intervening mechanism. 'Occult action at a distance' remains mysterious but is accepted because the laws, theories and models that derive from it fit so well with observations and experiments.

Many other fringe scientific ideas have survived scepticism, scorn and ridicule to become part of the mainstream. Philosophers who suggested that there was a connection between fiery streaks in the sky (meteors) and lumps of rock found on the ground (meteorites) were laughed at, and the concept that they might be rocks from outer space was literally heretical. Similarly, ball lightning, though often reported, was regarded by most scientists as a myth until physicist R.C. Jennison observed for himself a glowing sphere that passed through a passenger jet in which he was travelling one stormy night in 1963.

Tesla was a one-man fringe-science ideas factory, yet many of his apparently far-fetched claims were later vindicated. He claimed, for instance, to have created remote-control devices he called telautomatons; similar radio-controlled robots are now common. He claimed to have picked up messages from outer space; it is now thought possible he inadvertently created the first radio telescope. He claimed to be able to beam low-frequency electricity through space without wires, pumping energy into the ionosphere (one of the outer layers of the atmosphere); today the High Frequency Active Auroral Research Programme (HAARP) in Alaska does just this.

Beyond the fringe

Fringe science ideas can go in and out of fashion. The doctrine of panspermia, for instance, which holds that life on Earth was seeded by organisms carried here on asteroids, comets or interstellar dust, was popular in the 19th century but largely dismissed for most of the 20th, though it was championed by the popular scientist Fred Hoyle (*see* page 211). But the discovery of what appeared to be fossil bacteria in meteorite ALH84001, which had been blasted off the surface of Mars and subsequently landed on Earth, invested panspermia with fresh life.

Most fringe-science ideas, however, remain beyond the pale. Perpetual motion machines are known to be impossible thanks to the laws of thermodynamics (*see* page 16), and related concepts of 'free energy' have proved fertile ground for controversy. The claimed 1989 discovery of cold fusion, by Stanley Pons and Martin Fleischmann, caused considerable dispute and the field has been described by physicist David Goodstein as 'a pariah field, cast out by the scientific establishment. Between cold fusion and respectable science there is virtually no communication at all. In these circumstances, crackpots flourish.'

Flying car. The legendary Avrocar S/N 58-7055, a real-life flying saucer. This revolutionary experimental vehicle of the late 1950s was based on fringe technologies such as the Coanda effect and vertical takeoff and landing.

Future war. Avrocar marketing material, intended to sell the US military on the amazing potential of 'flying jeeps' for the battlefield of the future. Alas the technologies involved were too ambitious and could not be made to work. Eventually funding ran out and the project was shelved.

GLOSSARY

Attenuated vaccine: means of stimulating acquired immunity through exposure to a virus that is still whole but has been weakened

Big Bang: theory of creation of universe involving initial expansion from a single point

Biometrics: measurement of biological characteristics (such as height, IQ, etc.)

Calculus: mathematical system for calculating the gradients of curved lines and the areas under them

Catastrophism: theory that the Earth is acted upon by processes occurring in short, sudden, violent bursts ('catastrophes')

Cold fusion: nuclear fusion achieved at close to room temperature

Continuous creation: theoretical process by which new galaxies are formed to replace old ones that have died out

Copernican system: heliocentric model of Solar System proposed by Copernicus

Creationism: belief that God created the Earth and life

Crystallography: science of determining arrangement of atoms in a solid

DNA: deoxyribonucleic acid, complex helical molecule that carries genetic code

Dolichocephalic: a long, narrow skull shape

Doppler effect: phenomenon in which wavelength of a waveform changes as its source moves

Epistemology: the study of how knowledge is acquired and truth is discovered

Evolution: change in genetic material of organisms from one generation to the next

Experiment: a controlled manipulation of events, designed to produce observations that confirm or disconfirm one or more rival theories or hypotheses

Experimentum crucis: an experiment that shows the way for subsequent research

Fumarole: vent or crack in a volcano that gas and steam are emitted from

Galenism: school of medicine based on teachings of ancient Graeco-Roman physician Galen

Gene: a unit of genetic material that codes for a specific protein or protein component

Genome: genetic sequence

Geocentric: with the Earth at the centre

Geoheliocentric: with the Sun going round the Earth and everything else going round the Sun

Globuralism: belief that the Earth is a globe

Heliocentric: with the Sun at the centre

Hominid: great ape family that includes extant and extinct species of human and human ancestors

Hubble constant: number governing speed at which the universe is expanding

Hypothesis: informed speculation not properly tested/confirmed by experiment and observation

Industrial melanism: phenomenon whereby industrial pollution exerts selective pressure to produce melanistic (dark) forms of an organism

Intelligent Design: belief that evolution is directed by an intelligent designer

Inverse square law: law governing the strength of gravity, which decreases as the inverse square of distance

Junk DNA: DNA that does not code for genes

Killed vaccine: means of stimulating acquired immunity through exposure to fragments of a destroyed virus

K–T boundary: boundary between rock strata that marks the end of the Cretaceous and beginning of the Tertiary period; possible marker of K–T extinction event

Live vaccine: see attenuated vaccine

Long-period events: seismographic pattern warning of possible eruption

Microcephaly: disorder resulting in small head

Multiregionalism: theory that modern humans evolved from diverse, widespread species of human that migrated out of Africa relatively long ago

Nucleotide: structural unit of DNA; molecule composed of nitrogen, sugar and phosphate; comes in different varieties – the sequence of which constitutes the genetic code

Out of Africa: theory that modern humans evolved in Africa and only colonized the rest of the globe relatively recently

Palaeoanthropology: science of human origins

Palaeontology: study of prehistoric life

Panspermia: theory that life on Earth was 'seeded' by lifeforms that exist all through the universe

Pathogen: an organism that invades a host and causes harm, such as disease or poisoning

Planism: belief that the Earth is flat

Plate tectonics: theory that Earth's surface is made up of relatively thin plates of rock moving around on molten interior

Pseudoscience: theory or practice that has superficial similarity to science but lacks proper scientific methodology and/or evidence, especially testability

Psychoanalysis: school of psychotherapy originated by Sigmund Freud.

Ptolemaic system: geocentric model of Universe proposed by ancient Hellenic astronomer Ptolemy

Radio astronomy: astronomy conducted using radio waves instead of visible light

Radio telescope: telescope that gathers radio waves rather than visible light

Radiometry: measurement (particularly dating) through radioactivity

Red shift: phenomenon in which colour of light from an object moving away is shifted towards the red end of the spectrum

Retrovirus: virus that copies its genetic material into the DNA of a host cell

Scepticism: philosophical position involving questioning of authority and epistemology

Scholasticism: school of philosophy and epistemology based on ancient and religious authority

Seismology: study of earth tremors

Sequencing: process of determining sequence of nucleotides in a stretch of DNA

Speciation: evolution of new species

Spontaneous generation: development of organisms from inanimate matter

Teleology: study of design and purpose

Transference: process by which relationship templates acquired in early life are applied to other people/relationships

Uniformitarianism: theory that the Earth is acted upon by processes at a uniform pace

Vaccine: means of stimulating acquired immunity to confer protection against a viral pathogen

REFERENCES

Blazing a trail in the history of scientific feuds is the author Hal Hellmann, with his books *Great Feuds in Science*, *Great Feuds in Medicine* and *Great Feuds in Mathematics* (John Wiley & Sons).

Other useful general sources include *The Oxford Dictionary of National Biography*, the *Oxford Companion to Early Modern Science* and the 'Talk Origins Archive' website.

Part One: Earth Sciences

'An interview with Bernard Chouet', ESI Special Topics, March 2005, www.esi-topics.com/ volcanoes/interviews/BernardChouet.html, accessed 11 January 2009

Victoria Bruce, *No Apparent Danger The True Story of Volcanic Disaster at Galeras and Nevado del Ruiz* (HarperCollins Publishers) 2001

A. Hallam, *Great Geological Controversies* (Oxford University Press) 1983

Simon LeVay, *When Science Goes Wrong: Twelve Tales from the Dark Side of Discovery* (Plume) 2008

Matthew Rognstad, 'Lord Kelvin's Heat Loss Model as a Failed Scientific Clock', *The Age of the Earth,* www.usd.edu/esci/age/content/ failed_scientific_clocks/kelvin_cooling.html, accessed 18 January 2009

Tim Weiner, 'The Volcano Lover', *New York Times*, 15 April 2001

Stanley Williams and Fen Montaigne Fen, *Surviving Galeras* (Houghton Mifflin Company) 2001

Part Two: Evolution and Palaeobiology

Lawrence K. Altman and William J. Broad, 'Global Trend: More Science, More Fraud', *New York Times*, 20 December 2005

M. Brake and N. Hook, 'Darwin's Bulldog and the Time Machine', *Astrobiology Magazine*, 29 January 2007

Ronald W. Clark, *The Huxleys* (McGraw-Hill Book Co.) 1968

Joseph D'Agnese, 'Not Out of Africa', *Discover Magazine*, 1 August 2002

Richard Dawkins, *The Blind Watchmaker* (Penguin) 1990

Richard Dawkins, *The God Delusion* (Bantam Press) 2006

E. James Dixon, Bones, Boats and Bison: *Archaeology and the first colonization of the Americas* (University of New Mexico Press) 2000

G.B. Dobson, 'The Bone Wars', *Wyoming Tales and Trails*, www.wyomingtalesandtrails. com/ bonewars2.html, accessed 2 February 2009

Stephen Jay Gould, *The Panda's Thumb* (W.W. Norton & Co.) 1980

Bruce Grant, 'Sour grapes of wrath', *Science*, 9 August 2002

Paul Hancock and Brian J. Skinner, Eds, *The*

Oxford Companion to the Earth (Oxford University Press) 2000

Judith Hooper, *Of Moths and Men: Intrigue, Tragedy & the Peppered Moth* (Fourth Estate) 2002

Austen Ivereigh, 'Origin of the Specious', *America Magazine*, 12 January 2009

Donald C. Johanson and Maitland A. Edey, *Lucy: The Beginnings of Humankind* (Simon & Schuster) 1990

Horace Freeland Judson, *The Great Betrayal: Fraud in Science* (Harcourt) 2004

Joel Levy, *Poison: A Social History* (Quid Publishing) 2009

Roger Lewin, *Bones of Contention: Controversies in the Search for Human Origins* (University of Chicago Press) 1997

Richard Lewontin, 'Dishonesty in Science', *NY Review of Books*, Vol. 51, No. 18, 18 November 2004

Jim Mallet, 'The Peppered Moth: A black-and-white story after all', *Genetics Society News*, No. 50, January 2004

Robin McKie, 'I've got a bone to pick with you, say feuding dinosaur experts', *Observer*, 7 September 2003

Richard Milner, *The Encyclopedia of Evolution* (Facts on File) 1990

'Monte Verde Under Fire', *Archaeology*, www.archaeology.org/online/features/clovis/, 1999

Cole Morton, 'Baron Harries of Pentregarth: Can the bishop get the monkey off his back?', *Independent on Sunday*, 8 February 2009

Kevin Padian and Alan Gishlick, 'Books & Culture Corner: Of Moths and Men Revisited', *Christianity Today*, November 2002, http://www.christianitytoday.com/ct/2002/novemberweb-only/11-4-11.0.html

Keith M. Parsons, *The Great Dinosaur Controversy: A Guide to the Debates* (ABC-CLIO) 2004

Paul Raeburn, 'The Moth that Failed', *New York Times*, 25 August 2002

Arthur M. Shapiro, 'Paint it Black', *Evolution*, Vol. 56, No. 9, 2002

Smithsonian Museum, 'Human Ancestors Program', anthropology.si.edu/HumanOrigins/, accessed 20 January 2009

Chris Stringer, 'Piltdown's lessons for modern science', *BBC News*, 16 October 2006, http://news.bbc.co.uk/go/pr/fr/-/1/hi/sci/tech/6054656.stm

John Vidal, 'Bones of Contention', *Guardian*, 13 January 2005

David Rains Wallace, *The Bonehunters' Revenge* (Houghton Mifflin) 2001

John Noble Wilford, 'The Fossil Wars', *New York Times*, 7 November 1999

John Noble Wilford, 'The Leakeys: A Towering Reputation', *New York Times*, 30 October 1984

Matt Young, 'Moonshine: Why the Peppered Moth remains an icon of evolution', *TalkDesign.org*, www.talkdesign.org/faqs/moonshine.htm;, 11 February 2004

Part Three: Biology and Medicine

Colin Beavan, *Fingerprints: The Origins of Crime Detection and the Murder Case that Launched Forensic Science* (Hyperion) 2001

Sarah Boseley, 'Fall of the doctor who said his vitamins would cure Aids', *Guardian*, 12 September 2008

Sarah Boseley. 'Mbeki Aids denial "caused 300,000 deaths"', *Guardian*, 26 November 2008

Martin Brookes, *Extreme Measures: The Dark Visions*

and Bright Ideas of Francis Galton (Bloomsbury) 2004

Phyllida Brown, 'The strains of the HIV war', *New Scientist*, 25 May 1991

Michael Bulmer, Francis Galton: *Pioneer of Heredity and Biometry* (Johns Hopkins University Press) 2003

Dan Charles and David Concar, 'America hoists white flag in HIV war', *New Scientist*, 8 June 1991

Steve Connor, 'AIDS: Science stands on trial'; *New Scientist*, 12 February 1987

Lawrence Conrad, Michael Neve, Vivian Nutton, Roy Porter and Andrew Wear, *The Western Medical Tradition: 800 BC to AD 1800* (Cambridge University Press) 1995

John Crewdson, *Science Fictions: A Scientific Mystery, a Massive Coverup and the Dark Legacy of Robert Gallo* (Little, Brown) 2002

Roger French, *William Harvey's natural philosophy* (Cambridge University Press) 1994

Robert Gallo and Luc Montagnier, 'AIDS in 1988', *Scientific American*, 1988

Gerald L. Geison, *The private science of Louis Pasteur* (Princeton University Press) 1995

Nicholas Wright Gillham, *A Life of Sir Francis Galton: From African Exploration to the Birth of Eugenics* (Oxford University Press) 2001

James Gleick, *Isaac Newton* (Fourth Estate) 2003

Ben Goldacre, 'Don't dumb me down', *Guardian*, 8 September 2005

Ben Goldacre, 'Experts say new scientific evidence helpfully justifies massive pre-existing moral prejudice', *Guardian*, 18 April 2009

Nathan G. Hale Jr, 'Freud's Reich, the Psychiatric Establishment, and the Founding of the American Psychoanalytic Association: Professional Styles in Conflict', *Journal of the History of the Behavioural Sciences*; Vol. XV, No. 2, April 1979

Thomas Hayden, 'A genome milestone', *Newsweek*, 3 July 2000

John Henry, *Knowledge is Power: How Magic, the Government and an Apocalyptic Vision inspired Francis Bacon to Create Modern Science* (Icon Books) 2002

Lisa Jardine, *Ingenious Pursuits: Building the Scientific Revolution* (Doubleday) 1999

Horace Freeland Judson, *The Eighth Day of Creation: Makers of the Revolution in Biology* (Penguin) 1995

Walter Kaufmann, *Freud, Adler, and Jung – Discovering the Mind, Vol. 3*, (Transaction Publishers) 1992

Jeffrey Kluger, *Splendid Solution: Jonas Salk and the Conquest of Polio*, (Putnam) 2004

Jeanne Lenzer, 'AIDS "dissident" seeks redemption … and a cure for cancer', *Discover Magazine*, June 2008

Steven A. Lubitz, 'Early Reactions to Harvey's Circulation Theory', *Mount Sinai Journal of Medicine*, Vol. 71, No. 4, September 2004

Angela Matysiak, 'The Myth of Jonas Salk', *Technology Review*, July 2005

Paul A. Offit, *The Cutter Incident: How America's First Polio Vaccine Led to the Growing Vaccine Crisis* (Yale University Press) 2007

David Oshinsky, *Polio: An American Story* (Oxford University Press) 2005

Leslie Roberts, 'Controversial from the start', *Science*, 16 February 2001

Nils Roll-Hansen, 'Experimental Method and Spontaneous Generation: The Controversy between Pasteur and Pouchet, 1859–64', *Journal of the History of Medicine and Allied Sciences*, XXXIV(3), 1979

James Shreeve, *The Genome War* (Fawcett Books) 2005

'Sigmund Freud: Conflict and Culture', *Library of*

Congress, www.loc.gov/exhibits/freud/freud01.html

A.W. Sloan, 'William Harvey, Physician and Scientist', *South African Medical Journal*, August 1978

John Sulston and Georgina Ferry, *The Common Thread* (Bantam Press) 2002

Gavan Tredoux, '*Henry Faulds: the Invention of a Fingerprinter*', December 2003, http://galton.org

Craig Venter, *A Life Decoded: My life, my genome* (Viking) 2007

James D. Watson, *The Double Helix* (Penguin) 1968

Richard Westfall, *Never at Rest: A biography of Sir Isaac Newton* (Cambridge University Press) 1980

Michael Worboys, *Spreading Germs: Disease theories and medical practice in Britain, 1865–1900* (Cambridge University Press) 2000

Part Four: Physics, Astronomy and Maths

David Adam, 'Climate change sceptics bet $10,000 on cooler world', *Guardian*, 19 August 2005

John Aubrey (Ed. John Buchanan-Brown), *Brief Lives* (Penguin) 2000

Neil Baldwin, *Edison – Inventing the Century* (University of Chicago Press) 2001

Margaret Cheney, *Tesla: Man Out of Time* (Prentice-Hall Inc.) 1981

Owen Gingerich and Robert S. Westman. *The Wittich connection: conflict and priority in late sixteenth-century cosmology* (Diane Publishing) 1988

James Glanz, 'What Fuels Progress in Science? Sometimes, a Feud', *New York Times*, 14 September 1999

David Goodstein, 'Whatever Happened to Cold Fusion?', *California Institute Technology*, www.its.caltech.edu/~dg/fusion_art.html, accessed 28 March 2009

N. Jardine, D. Launert, A. Segonds, A. Mosley and K. Tybjerg, 'Tycho v. Ursus', *Journal for the History of Astronomy*, Vol. 36, Part 1, No. 122, 2005

Kevin Kelly, 'A Brief History of Betting on the Future', *Wired*, 10.05, May 2002

Kristine Larsen, *Stephen Hawking: a biography* (Greenwood Publishing Group) 2005

Robert Lomas, *The Man Who Invented the Twentieth Century* (Headline) 1999

Tom McNichol, *AC/DC: the savage tale of the first standards war* (John Wiley and Sons) 2006

John Michell, *Eccentric Lives and Peculiar Notions* (Thames and Hudson) 1984

Simon Mitton, 'A truly stellar career that ended with a big bang', *Times Higher Education*, 15 April 2005

Richard Owen and Sarah Delaney, 'Vatican recants with a statue of Galileo', *The Times*, 4 March 2008

Peter Popham, 'Science bows to theology as the Pope dismantles Vatican observatory', *Independent*, 4 January 2008

Edward Rosen, *Three Imperial Mathematicians: Kepler Trapped Between Tycho Brahe and Ursus* (Abaris Books) 1986

Marc Seifer, *Wizard: The life and times of Nikola Tesla* (Birch Lane Press) 1996

Charles H. Smith, 'Letters to the Editor Concerning the Bedford Canal "Flat Earth" Experiment', *The Alfred Russel Wallace Page*, February 2009, www.wku.edu/~smithch/wallace/S162-163.htm

Charles Webster, *From Paracelsus to Newton: Magic and the Making of Modern Science* (Cambridge University Press) 1982

Michael White, *Isaac Newton: The last sorcerer* (Fourth Estate) 1998

INDEX

ACKNOWLEDGEMENTS

The author and publishers would like to extend their thanks to the following copyright holders for allowing their photographs to be used in this book.

(GI = Getty Images; C = Corbis; TF = Top Foto; GC = The Granger Collection, New York; SPL = Science Photo Library; IS = iStock)

p3: Thomas Brostrom/IS; pp4-5: GC; p9: Stefano Bianchetti/C; p11: GC; pp12-13: Blue Earth: NASA; p14: Kelvin: GI, Lyell: GI; p16: HIP/TF; p20: Tony Colter/IS; p22: Wegener: Alfred-Wegener-Institut, Jeffreys: Royal Astronomical Society/SPL; p24: Alfred-Wegener-Institut; p25: Alfred-Wegener-Institut; p27: Alfred-Wegener-Institut; p31: Bernard Chouet; p35: AFP/GI; pp36-37: Fossil Ammonite: Heiko Grossman/IS; p38: Huxley: Humanities and Social Sciences Library/New York Public Library/SPL, Wilberforce: GC; p41: Science Source/SPL; p42: Darwin: Carina Lochner/IS; p44: The British Library/HIP/TF; p47: Dusko Despotovic/C; p48: Cope: GC, Marsh: Science, Industry and Business Library/New York Public Library/SPL; p50: World History Archive/TF; p51: Dinosaur: breckeni/IS; p52: wyomingtalesandtrails.com; p56 Leakey: Peter Arnold, Inc./Alamy, Johanson: John Reader/SPL; p59: John Reader/SPL; p60: Des Bartlett/SPL; p64: Newspix/News Ltd/3rd Party Managed Reproduction & Supply Rights; p68: Time & Life Pictures/GI; p71: Claude Nuridsany & Marie Perennou/SPL; p72: Topham Picturepoint; p73: GC/TF; p75: GI; p76: Bettmann/C; p79: breckeni/IS; p80: Clovis Point: Apollo Design Studios/IS; p81: Emmanuel Laurent/Eurelios/SPL; p84: NASA/courtesy of nasaimages.org; p89: Equinox Graphics/SPL; p90: Newspix/News Ltd/3rd Party Managed Reproduction & Supply Rights; pp92-93: Osuleo/IS; p95: GC; p96: GC; p97: Bettmann/C; p98: Bettmann/C; p100: SPL; p104: GC; p107: Chicken and Egg: Ivonne Wierink-van Wetten/IS; p108: Faulds: Gavan Tredoux, Galton: Gavan Tredoux; p109: Fingerprint: Hanquan Chen/IS; p111: GC; p112: Topham Picturepoint; p113: Gavan Tredoux; p115: Gavan Tredoux; p116: Adler: GC; p118: GI; p120: Freud Museum, London; p123: Freud Museum, London; p124: Jung: SPL; p127: GC; p128: Freud Museum, London; p130: Sabin: Bettmann/C, Salk: Bettmann/C; p132: Time & Life Pictures/GI; p133: GI; p134: GC, CDC/PHIL/C; p137: GI; p138: Bettmann/C; p139: Bettmann/C; p140: Franklin: From the Jewish Chronicle Archive/HIP/TF, Wilkins: Topham Picturepoint; p141: A. Barrington Brown/SPL; p143: Frances Evelegh/SPL; p144: Topham Picturepoint; p146: Montagnier: SPL, Gallo: Bettmann/C; p148: GI; p151: GI; p153: AFP/GI; p154: Congressional Quarterly/GI; p155: GI; p156: Venter: Jurgen Frank/C Outline, Collins: Brooks Kraft/C; p160: Ron Sachs/CNP/Sygma/C; p161: alohaspirit/IS; pp162-163: Constellation of Ophiucus, Plate 22 from Flamsteed's Atlas Coelestis: Stapleton Historical Collection/HIP/TF; p165: GC/TF; p166: Stefano Bianchetti/C; p168: David MacLurg/IS; p169: GC/TF; p170: GC; p171: GC; p173: GC; p174: Roger-Viollet/TF; p175: Time & Life Pictures/GI; p176: Newton: The Gallery Collection/C, Flamsteed: GC; p178: Jim Sugar/C; p179: Royal Astronomical Society/SPL; p180: GC; p182: Newton: The Gallery Collection/C, Leibniz: GC; p184: GC/TF; p185: GC; p191: Bob Caddick/Alamy; p193: Hulton Archive/IS; GC; p194: SPL; p196: Tesla: GC, Edison: GC; p197: Jonathan Sweet/IS; p198: Bettmann/C; p200: GC; p202: Niday Picture Library/Alamy; p204: Ryle: Bettmann/C, Hoyle: Hulton-Deutsch Collection/C; p206: Topham Picturepoint; p210: Clayton Hansen/IS.